DNA Pioneer
J. Herbert Taylor
1916–1998

J. Herbert Taylor
DNA Pioneer

**Shirley Taylor
and
Joan T. Hare**
Florida State University

Published by Florida State University, Tallahassee, Florida

© 2005, Florida State University, All rights reserved.

For additional copies contact: College of Arts and Sciences, Director of Development, 110 Longmire Hall, Ivy Way, Florida State University, Tallahassee, Florida 32306-1280. Phone: 850-644-9324

ISBN 0-9766281-0-4

Published in the United States of America by Florida State University, College of Arts and Sciences.

Reprinted material is quoted with permission.

Foreword excerpted in part from J. Hare, *American Society for Cell Biology Newsletter,* 1999.
The font in this book is Adobe Garamond.
Book design and typeset by Graphic Edge, Inc., Tallahassee, Florida.
Cover art by Charles Badland, Florida State University
Printed by Graphateria, Tallahassee, Florida.

Shirley Taylor, December 1920 – August 2004

In the final years of Shirley Taylor's remarkable life, she poured her energy into the creation of this manuscript. For Shirley, it was much more than just a chronicling of events; it was the reliving of the dreams and rich history she experienced with Herb Taylor in a marriage spanning nearly sixty years. This volume is a testament to her lifelong love and commitment.

Contents

Preface	ix
Foreword	xi
Acknowledgements	xiii
Announcing the J. Herbert Taylor Distinguished Professorship in Molecular Biology	xvi

Part I: Memoir

	Chapter 1.	Life of Discovery	1
	Chapter 2.	The Experiment Heard Round the World	29
	Chapter 3.	Fame Abroad	51
	Chapter 4.	Models of Chromosomes	69
	Chapter 5.	DNA Replication—From the Chromosome to the DNA Molecule	85
	Chapter 6.	Living in the Community—of Science, the University, and Tallahassee	97

Part II: Tributes

Memories of Herb Taylor as Father, Brother, Brother-In-Law	111
Memories from the Community	119
Memories of Herb Taylor as Scientific Colleague	133
Memories of Herb Taylor as Mentor —from Scientists who Trained in His Laboratory	149

Appendix:

List of friends, colleagues, students and postdoctoral fellows of Dr. J. Herbert Taylor	199
J. Herbert Taylor, Curriculum Vitae	209

Preface

A lifetime viewed retrospectively is often surprising in the twists and turns that bring an individual to a specific time and place. Herb Taylor's nomination to the National Academy of Sciences in 1977 was the impetus for him to consider and to record his experiences in the context of history, to begin to trace his path. Later, following his retirement in 1995, he received several requests to record his remarkable experiences in essay form, a process that led to reflection and analysis of his time and place.

In Herb Taylor's writings about his life, he referred often to the Robert Frost poem "The Path Not Taken." The poem's most remembered line, "I chose the one less traveled by / and that has made all the difference," held a special meaning for Herb Taylor as he considered the path he had taken. Herb Taylor's path led him to the world stage as a key actor in the drama of the discovery of the structure of DNA. How he got there and what happened after he arrived is the story of this memoir.

When the news of Herb Taylor's death at the end of 1998 was sent to his colleagues and graduates by his former student, Joan Hare, scores of messages came back to Joan and Herb's wife Shirley. It became clear to them that the wealth and warmth of memories of Herb and times past recounted in the letters ought to be shared with all the senders. As Joan and Shirley assembled the letters, the problem of production and distribution of the collection loomed large.

Meanwhile Shirley, with the support and encouragement of the university provost, began a five-year development plan to endow a J. Herbert Taylor Chair of Molecular Biology. Joan got in touch with Taylor's colleagues and graduates again to invite their contributions to help support the endowment. Time was passing, and a new idea surfaced—perhaps the book could be used as a fundraising tool. If so, Mary Esther Gaulden suggested, the book would need a brief introduction that covered the life and work of J. Herbert Taylor.

Impressed by E. O. Wilson's scientific memoir *The Naturalist,* Shirley and Joan decided to undertake an even more ambitious project, to produce a book that would include not only those messages of remembrance but also a personal and research memoir to trace how Herbert

ix

Taylor became "hooked on research"—and never stopped. Writing and compiling this memoir has taken them more than two years—searching out relevant materials, selecting which of all the interesting pieces to include, finding a structure, and getting all the details right.

Shirley produced the personal-life memoir portions, based on sections from Herb's own unpublished "My Autobiography: The Road from Drain (Texas)" and portions of letters the two exchanged while Herb was in the South Pacific during World War II. Joan has painstakingly traced the path of Taylor's DNA research through the years, on the basis of his own accounts published and unpublished, his scientific papers, and her years of work with him.

Our work to create this book was challenging, at times frustrating, but the result is its own reward. For all of you who will peruse this volume, we hope that you may uncover a cherished memory, learn, laugh, and perhaps find a bit of inspiration. Please enjoy following along Herb's path less traveled.

— Shirley Taylor and Joan Hare, July 2004

Foreword

James Herbert Taylor was a pioneer in molecular genetics. His career spanned more than 50 years, during a time of phenomenal change in the field of genetics, from before the structure of DNA was proposed to the isolation of eukaryotic origins of DNA replication. His groundbreaking contributions to the knowledge of chromosome structure and reproduction produced many methods now standard in genetics and changed the way many aspects of chromosomes were presented in genetics textbooks. Taylor's early work in the mid 1950's helped to prove that DNA was the genetic material of the chromosome, by demonstrating that a radioactive label incorporated into DNA could be visualized in the autoradiograms of chromosomes. In similar experiments in the germ cells of plants, Taylor made the critical genetic observation that most DNA synthesis occurs prior to the earliest stages of meiosis, a finding that forced the models of genetic exchange to be re-examined.

In 1956, in work that almost instantly caught the attention of the entire scientific world, he clearly demonstrated that the DNA of eukaryotic chromosomes was synthesized and segregated in a semiconservative fashion, providing the first and some of the most compelling evidence in favor of the still-controversial model of DNA structure and synthesis proposed by Watson and Crick just two years earlier. Equally important but not realized for another two or more years was the controversial conclusion from this work and his later work, that a chromosome was composed not of multiple copies, as was then believed, but of a single and continuous strand of DNA. The evidence lay in his demonstration that sister chromatid exchanges were not a rare esoteric phenomenon but a common event in mitosis and that the occurrence of twin or of single exchanges revealed in which round of replication an exchange event had occurred. Taylor made the simple but brilliant deduction that a fit to a predicted ratio could confirm a second feature of the Watson-Crick model. Analyzing data from his first several experiments with the tritium label he confirmed the correctness of this second feature of the Watson-Crick model—the DNA molecule was composed of two strands that were replicated and separated as if they were antiparallel. The discovery of sister chromatid exchanges lead also to the development of a cytogenetic method that is still viewed today as one of the best *in vitro* methods for predicting the carcinogenicity of compounds and measuring genotoxicity.

The autoradiographic techniques Herb Taylor developed with his scientist wife, Shirley Hoover Taylor, and his pioneering development and use of a radioactive DNA precursor, tritiated thymidine, opened to geneticists an entirely new area of investigation into chromosome duplication and division. Earlier efforts with P^{32} lacked the necessary resolution and specificity. Tritium had never before been used in a DNA precursor; Taylor and his coworkers made their own—the first tritiated thymidine. Taylor's highly successful demonstrations of the power of the tritiated precursor, thymidine, in conjunction with autoradiograms revolutionized DNA synthesis experiments. His success led commercial companies to quickly develop tritiated thymidine and to make it available for scientific use. Fifty years later, these techniques are still prominent in several areas of biochemistry, medicine and molecular biology.

Using these techniques, he discovered an ordered temporal pattern of DNA replication in mammalian cells including regions of chromosomes which replicated late—most notably one entire X chromosome, thus beginning an area of study correlating gene activity, condensation of chromosomes and time of replication. Over the next several years, he continued to make a series of significant discoveries about important details of DNA replication in eukaryotic chromosomes. Almost 10 years after he first captured the scientific world's attention, he completed the much more difficult experiment that he had first dreamed of—demonstrating that genetic recombination in the germ cells of plants involved a physical exchange of preexisting DNA.

Herb Taylor had a deep and abiding curiosity about the organization of chromosomes and the mechanisms of chromosome reproduction. It was his fascination with how things worked—the mechanisms of cell reproduction and division, the intricate ballet of chromosome pairing and genetic exchange—that guided his entire career. Herb Taylor changed the way many geneticists thought about some of the central issues of his time. His influence on the field came also though his exposition of models. Not all were correct, but all stimulated the thought of other scientists who later proved, disproved, or improved those models. Taylor was never intimidated by new ideas. The process of discovery had to include the willingness to be proved wrong. Taylor later wrote, "Nothing is so clear as a *proven* scheme or theory, but in the developmental stages there is often a bewildering array of experiments and hypotheses to sort through and few guideposts for the adventurer into the unknown. That remains the fascination and adventure of science—forging into the unknown where no one has been before."

Acknowledgements

I appreciate the help by these friends who read my memoir section and offered helpful comments toward shaping its final form: Temple and Betsy Neuman, Bill Haut, Bob Hopkins, Joyce Anthony, Mary Esther Gaulden. My son and daughters, Michael and Lucy Taylor and Lynne Taylor Ireland, have been supportive and patiently provided help and useful advice throughout. I especially thank Michael and Diane Taylor, my son and daughter-in-law, for their continuing and tireless assistance, especially in the final stages.

I am most grateful to Florida State University Provost and friend Lawrence G. Abele for his encouragement and continuing support in helping me initiate and develop the J. Herbert Taylor Chair in Molecular Biology and the production of this book. I especially appreciate the thoughtful assistance of good friend and Blandy Farm graduate, Dr. Mary Esther Gaulden, in developing the concept and shaping goals for this ongoing memorial to Herbert Taylor.

Thanks to Dean Foss, his staff, and the FSU Foundation for assistance in tracking funding of the Taylor Chair in Molecular Biology. For editorial assistance, Anne B. Thistle, Department of Biological Science, FSU, has been most helpful, as have Jane Houle and Brian Zeiger, editors for the book design and layout.

And only with Joan Hare's initial encouragement and spadework, and her continuing commitment day to day, could this book have been completed. Finally to all of you whose messages appear in this book, I express my deep gratitude for the immediate and spontaneous responses that remain so comforting to me.

— Shirley Taylor, Williamstown, Massachusetts, July 2004

Many thanks to Kurt Hofer and Sheldon Wolff who offered valuable input to this manuscript in its final stages.

My thanks go to Hank Bass, FSU Department of Biological Science, for reading early meiosis papers and discussing the meiosis section with me; to my husband, Jed Dillard, for reading the entire memoir and offering his opinion from the layman's viewpoint; and to Jane Houle of Graphic Edge, Tallahassee, FL whose design guidance while the book was still in the development stage was invaluable. Her efforts have given the book a much more readable, and I believe, visually pleasing form. A second thanks to Anne Thistle, our ever-vigilant editor, who as Shirley once exclaimed, "knows punctuation rules that stagger me!" Thanks also to Steve Leukanech of FSU Chemistry Department for the reproduction of photomicrographs and figures, sometimes with short notice! And a special thanks to Michael Taylor who became the closer for the entire book project, making sure all those final details got attended to. Diane Taylor did a wonderful job of collecting and reproducing family photographs for the memorial gathering. We appreciate her efforts in reproducing these and other family photographs for this book. Together she and Shirley spent many hours selecting, scanning, and placing these family treasures.

We thank Terry Ashley, Yale University, who originally suggested that we compile and distribute the messages and tributes we received—that was the beginning of this book. Their comments became both the impetus to produce this book and the background material for it.

What later began as a brief recounting of Taylor's scientific accomplishments as an introduction for the letters became so much more when Shirley began to contribute the stories of the couple's and family's life. This additional detail has greatly enriched the biography. The wartime letters between Herb and his then fiancée, Shirley and travel adventures of their later years told in the couple's Christmas newsletters add richness to the story, possible only through her generosity in sharing them with us. Her commitment to this book and to the Taylor Chair has been a driving force in her life in the last four years and in mine. She has always been the one who kept any of us from straying too far from our intended path of writing a book, or funding a Chair, or establishing a Chair. I am grateful for both her patience and her impatience, seemingly at the right times. This book has been a labor of love for me of both Herb Taylor and Shirley Taylor and a testament of enduring respect for them both -mentors in science, and in life!

— Joan Hare, Tallahassee, Florida, July 2004

Note added following Shirley's Death:
As her health began to fail at the end, her commitment to this book never wavered — Shirley's primary concern was to finish this book. Her strong commitment, her deep belief in this project inspired me. And more, it challenged me to work harder than I knew I could in that last stretch. And though she didn't live to hold the printed book in her hands, she did live to see the designer's proofs. She left the rest to us. With the help of many whose names appear earlier, I have kept the promise.

—Joan Hare, December 2004

Announcing the J. Herbert Taylor Distinguished Professorship in Molecular Biology

The Department of Biological Science at The Florida State University is most pleased to announce that it will be home to the J. Herbert Taylor Distinguished Professorship.

This endowed faculty position will serve as a lasting tribute to one of the department's most outstanding and distinguished faculty members. Current and future faculty and students in the Department will be honored to have the Department forever associated with such a pioneering, dedicated, and creative scientist. It is particularly apropos to announce this position in the present volume, which does so much to capture the spirit of the man as well as the nature of his scientific contributions.

"The J. Herbert Taylor Distinguished Professorship in Molecular Biology is established to promote experimental and theoretical research that crosses disciplines in an effort to elucidate the complexities of gene action that are involved in all aspects of living systems. Florida State University will bring a proven scholarly researcher of international reputation who will attract students and recruit faculty to utilize disciplinary convergence with the overall goal of solving complex problems of gene action and its many ramifications in organisms. The holder of this professorship will lead and encourage both students and faculty to participate in interdisciplinary studies that advance the knowledge of the cell—reproduction, differentiation, and function—and the control thereof, research topics that were at the core of Dr. Taylor's interests and work."

We wish to take the opportunity publicly to thank those who have made this endowment possible; in particular we express deep gratitude to Dr. Shirley Taylor.

Timothy S. Moerland
Professor and Chair
Department of Biological Science
Florida State University

◇

Further contributions to the endowed fund are welcome and may be sent to the Development Office, College of Arts & Sciences.

College of Arts & Sciences
Director of Development (Ms. Nancy Smilowitz)
Longmire Hall, Ivy Way
Florida State University
Tallahassee, FL 32306-1280

Contributions should be made out to the Florida State University Foundation, with a written notation that the gift is intended for the J. Herbert Taylor Endowed Fund.

◇

Selection Committee

A Selection Committee of highly qualified scientists and academics has been assembled to conduct a nationwide search and develop a list of potential candidates. The committee is composed of three members of the Florida State University faculty of varying lengths of experience and two distinguished scientists from other institutions, who have been recommended by the donor. From this list of candidates, the President of the University will name the J. Herbert Taylor Distinguished Professor.

In addition to selecting candidates consistent with the purpose and intent of the Distinguished Professorship, the committee will continue to serve in an advisory role and will be kept informed annually of the program's progress.

Current Members of the Selection Committee

Timothy S. Moerland (Selection Committee Chair)
Professor and Chair
Department of Biological Science
Florida State University

Hank W. Bass
Assistant Professor of Biological Science
Department of Biological Science
Florida State University

Joseph Gall
Adjunct Professor
Department of Biology and Department of Embryology
Carnegie Institution of Washington

Kurt Hofer
Retired 2003 from Institute of Molecular Biophysics
Florida State University

William F. Marzluff
Kenan Professor of Biochemistry and Biophysics
Executive Associate Dean for Research
University of North Carolina School of Medicine

From the Selection Committee

Tim Moerland
"The J. Herbert Taylor Distinguished Professorship ensures that Herb's vision and passion will be part of the Department of Biological Science and Florida State University well into the future. I am deeply honored to have the opportunity to help remember him in this way."

Hank Bass

"Serving on the J. Herbert Taylor Distinguished Professorship search committee is a high honor and great delight. His works and writings have inspired me with the passion and courage to pursue some of life's great mysteries, the biology of DNA and the behavior of genetic material. The professorship honors Dr. Taylor's contributions to science, his time at Florida State University, and his keen focus on fundamental questions of genetics. Dr. Taylor would have been pleased to learn that some of his questions have finally been answered, but equally delighted to learn that others remain open for exploration by future scientists."

Joe Gall

"Herb Taylor was a good friend and colleague, whose work on chromosome structure and function was a great inspiration for me, especially when I first started my scientific career."

Kurt Hofer

"I have the utmost respect and admiration for Herb's accomplishments as a scientist and I believe that the J. Herbert Taylor Distinguished Professorship is a fitting tribute to this remarkable man. But while this is true, my real motivation for volunteering to work on the Taylor Professor Search Committee is more personal. Herb was both my mentor and my close personal friend. He helped me a lot and he never asked for anything in return. Making a special effort to find a worthy candidate for the Taylor Professorship is my way of thanking Herb for what he did for me."

Bill Marzluff

"I met Herb Taylor when I joined the faculty at Florida State University in 1974, although like every other student of that generation I had read his book with the collected seminal papers in molecular genetics. During the 15 years I was on the faculty with Herb, I served on the committee of several of his graduate students when very exciting work on DNA replication and the structure of chromosomal DNA was going on. What sticks out about Herb is not only was he a superb scientist, but he was a wonderful person, one who was committed to science and training young scientists. I look forward to the recruitment of an outstanding scientist to the endowed chair who will bring both the scientific and personal qualities that will keep Herb's memory alive at Florida State."

The Road Not Taken

Two roads diverged in a yellow wood,
And sorry I could not travel both
And be one traveler, long I stood
And looked down one as far as I could
To where it bent in the undergrowth;

Then took the other, as just as fair,
And having perhaps the better claim,
Because it was grassy and wanted wear;
Though as for that the passing there
Had worn them really about the same,

And both that morning equally lay
In leaves no step had trodden black.
Oh, I kept the first for another day!
Yet knowing how way leads on to way,
I doubted if I should ever come back.

I shall be telling this with a sigh
Somewhere ages and ages hence:
Two roads diverged in a wood, and I—
I took the one less traveled by,
And that has made all the difference.

Robert Frost

Part I
Memoir

Herbert Taylor discovered research at a time in the twentieth century when research could be done by scientists as individuals. And in a time when finding one's research specialty was itself a process of discovery. In his writings he recounted how he came to be a biologist, cytogeneticist, and molecular biologist and in all of these an avid researcher ... and how he enjoyed life.

Chapter 1

Life of Discovery

"When I was born, my parents lived on a farm in the Cotton Belt of north central Texas, but we soon moved into Dallas for about a year, where my father operated a taxi service. In 1918 we moved to south central Oklahoma, where I grew up, attended school, and graduated from a small state college. When we arrived in Oklahoma, the new state was just emerging from the primitive conditions that characterized its transition from Indian Territory to statehood and the arrival of many more white settlers. Many of my boyhood acquaintances and friends were Indians, and several of my grade-school teachers were Choctaws. Ours was a farming community, and many wild areas remained for hunting and fishing. The free range for cattle was being displaced by small farmers who came to occupy the land. In this environment, educa-

J. Herbert Taylor (left, standing) at 20 years of age, with brothers and sister, Bennington, Oklahoma, 1936

tional opportunities were limited, but my parents were determined to give their four children opportunities they had not had; neither of them had attended college. Their hopes were partially frustrated and delayed, however, by the depression years, crop failures, and the dust bowl conditions just west and north of us. These conditions contributed to a severe financial collapse of that part of the country" (unpublished memoir, 1997).

Herb's younger brother, Tom Dotson Taylor, recalls this decisive point in 1933, when they were operating a "binder," the large farm machine for cutting and binding oats or other grains.

"We were binding a late, overgrown field of oats in a river bottom in southeastern Oklahoma near the Red River. It was late August, humidity near 100 percent, temperature around 100 degrees. A wet summer had caused the oats to grow shoulder high and the binder would stall when we hit spots of wet soil. The binder was a heavy load for the four mules to pull, especially in the tall oats and in wet soil. The machinery of the binder was run off a 'bull wheel', a large wheel whose chain gear hooked to all the binder's operating equipment. When we hit a soft wet spot the bull wheel would slip and all the moving canvas and cutting blade would jam up with hay. We would then have to pull all the jammed up hay out of the equipment and finally turn the gears with a hand crank to clear hay out completely. It took both of us to get the crank started and then several minutes of turning. Every fifteen minutes we were forced to repeat these operations.

"When we had stopped for perhaps the tenth time that afternoon, we were standing back to catch our breath and cool off a bit while we studied the sky, wondering if we might get an afternoon thunderstorm. I noticed that Herb had stepped off a bit, looking over the whole situation in a careful study. I thought maybe he was considering whether we should pull out to let the ground dry more. Instead he looked square at me and proclaimed, "I don't know about you, T.D., but I am not going to spend any more of my life doing this!" We pulled out of that field, secured the equipment, unhooked the mules, and headed for the barn just ahead of a thunderstorm.

"About a week later Herb was off to the CCC Camp, where he spent the next year. He enrolled in college the next fall, and sure enough—he never returned to that river-bottom oat field or any part of farm life again. He found a much more suitable calling."

After these three post–high school years of farming and Civilian Conservation Corps experience in eastern Oklahoma, Taylor began study at nearby Southeastern State College. He had been inspired by the CCC director to pursue an engineering–bridge building career and wanted to attend an engineering school, but that would be farther away and more costly than he could afford. Looking back in 1991, Herb recalled:

"I began by majoring in math and the physical sciences, hoping that I could transfer to an engineering school later, but I soon realized that I would have to find a job, and the best opportunities were available for teachers of biology. I changed to a major in biology with a minor in math, and I took Latin for my language requirements. I soon developed a deep interest in biology, especially taxonomy and natural history, for I had grown up in close contact with the wilderness. I had collected insects and small animals as a hobby or to keep as pets, but it had not occurred to me to pursue such a study for a lifetime. I was also intrigued by both chemistry and physics, and for a time I considered changing into physics" (unpublished memoir, 1997).

◇

Hooked on Research

Taylor: "In 1939 I finished my B.S. degree. Dr. Walter L. Blaine, one of the two biology professors at the college, suggested that I go on to graduate school, but I felt that I should use my teacher's training and try that occupation for a while … and I needed to earn some money. I obtained a position to teach sciences—biology, chemistry, and physics—in a high school in southeastern Oklahoma. I knew before half the year had passed that I had not found the type of work I would like to continue.…

"[But] I wondered what I could do in graduate school. I was told that original work was required, but when I read the books, it looked as if everything was already known. [It seemed] All the questions were answered if one looked in the proper book. I had no contact with anyone who knew enough to give me a view of the unknown, the many unanswered questions in science.

"With Dr. Blaine's assistance, I obtained a small fellowship stipend and admission to the Department of Botany and Bacteriology at the University of Oklahoma. My course in plant physiology was taught by O. J. Eigsti, who had recently discovered how to produce plants with double the normal chromosome number by the use of a drug called colchicine. By arresting cell division and then allowing recovery and division of the arrested cells as the drug concentration was reduced, tetraploids could be produced. I produced some tetraploids and began some comparative studies of physiology of diploids and polyploids[1] and measuring their response to water in the soil and atmosphere. I realized that I was learning new things about plants that had never existed before, or at least were not known by scientists. These were minor matters that might not be very significant or useful, but they were new. I had found the world of research, the unknown area in which there were ways of finding answers that no one else knew. Here was my world—no other would ever rival it. Now I knew what research was like and I was hooked. Life for me would never be the same again, for I had a glimpse of the unknown. I would need to learn how to look for the unknown and to acquire the knowledge to recognize the significant facts when revealed.

"Years later I was unwilling to change my research in a way that would meet the practical objectives of the Atomic Energy Commission or any other applied scientific objective. I held to that principle throughout my research career. I would pursue fundamental scientific questions and go wherever that led me. I have been lucky in that I could take advantage of some of the funds made available for practical projects and bend the objectives to fit the fundamental research I have wanted to do. I came into research at a fortunate time for me, in that the great surge to advance large practical projects left some portion that could be used to pursue the fundamental unanswered questions of science as long as the questions asked and pursued had a chance of gaining insight into the nature of the universe and its creatures regardless of the perceived value at a particular time."

... "I had two job opportunities after finishing a master's degree [in Oklahoma] in just nine months. I could work in Illinois in hybrid seed-corn research or go to [the University of] Virginia, where I had been offered a fellowship [at Blandy Farm] to study cytogenetics.[2] I spent that summer in the corn fields and learned some of the techniques of producing and the propaganda of selling hybrid corn, but I never wavered from

---

[1] *polyploid: having multiple sets of chromosomes or more than the usual diploid set of chromosomes.*
[2] *cytogenetics: the study of heredity and variation using cytology and genetics.*

my determination to return to graduate school and the lure of research into the truly unknown" (unpublished memoir, 1997).

Accepting the Blandy Farm fellowship, Herb headed east to continue graduate study and research in Virginia. Not only did he find there a greener world with mountains and more large trees, but the University of Virginia with all its history and traditions was totally unlike the midwest of Herb's previous experience. Mr. Jefferson's university had many buildings that he had originally designed and were still in use. There was never a "campus" but what was and still is known as The Grounds. The college was exclusively for males, and women were admitted to only a few of the graduate-school departments. The fifty or so coeds (from the entire university!) had their own Dean of Women and one cozy room set aside for daily relaxation and lunching. First-year men, even graduate students, wore hats and coats to class and on The Grounds. Virginia, with its history, tradition, and formality, was altogether different from the informal and newer midwestern world to which Herb had been accustomed, and adjusting required some effort. All student exams and behavior were governed by the honor system, and all disciplinary action was decided by their peers, as judged in a student court.

The university's Miller School of Biology was of modest size compared to departments today. The seven-professor faculty taught cytology,[3] cytotaxonomy,[4] parasitology, comparative anatomy, evolution, general zoology, and general botany. There were no more than twelve or fifteen graduate students, congenially working long hours till midnight, except on weekends. Most of the professors were avid researchers who also spent considerable time interacting one on one with their students. Herb later recalled:

Taylor at the University of Virginia's Blandy farm, 1942

"I arrived in Charlottesville about mid-September, 1941, and made my way to the biology building (The Miller School of Biology) to see Dr. Orland E. White, my sponsor and advisor. I was not prepared for this bear

---
[3]*cytology: the study of the structure, function, multiplication, pathology, and life history of cells.*
[4]*cytotaxonomy: study of the relationships and classifications of organisms using both classical systematic techniques and comparative studies of chromosomes.*

of a man with huge, bushy eye brows and hair growing out of his ears plus a little goatee. His rumpled suit was in marked contrast to Dr. Cross or to Dr. Milton Hopkins [my professors in Oklahoma]. I knew Dr. White was a Harvard graduate, but he had neither the accent nor the manners of the few Harvard men I had known. I later learned that he had grown up in South Dakota, came to Harvard for graduate work with Dr. E. M. East, and before coming to Virginia he had spent ten years at the Brooklyn Botanical Garden, part of that time as a plant explorer in the Amazon basin. He attended seminars at Columbia University, where he met Thomas Hunt Morgan and A. H. Sturtevant, both of whom became famous geneticists. He had moved to Virginia in the early 1930's to head a research station, The Blandy Experimental Farm, located in northwest Virginia about 60 miles from Washington, D.C. Dr. White liked to initiate his students with some outlandish story to publicize their arrival. He told Dr. Kepner, the very proper Virginia gentleman, that I arrived from the west on horseback. He even elaborated on this by saying that I attempted to ride my horse through the door into the Miller School. With this type of beginning I was uncertain of my first term's prospects.

"My first course in genetics was different from anything I had expected. Any formal genetics that I learned was in the lab or outside the classes.

(top) Blandy Farm lab and living space, University of Virginia, 1943

(right) Orland E. White at Blandy Farm, University of Virginia, 1929

Orland E. White at home (Taylor's professor at Blandy Farm), 1965

Dr. White taught genetics based on historical developments and the personalities who developed the science, many of whom he knew personally. Genetics was a new science that was named by Bateson after 1900 when the work of Mendel was rediscovered. White's lectures provided a wonderful perspective but left me deficient in the fundamentals of genetics. Dr. White was not the precise lecturer that I had encountered in Dr. Cross. White was a great story teller, and he would act out some of the dramatic advances, especially those announced at national or international meetings. He also provided the lineages of students, especially from American universities. The great labs were E. M. East and Castle at Harvard, R. A. Emerson at Cornell and T. H. Morgan, first at Columbia and later at Cal Tech in Pasadena, California.

"Dr. Kepner taught the course in invertebrate zoology. Since I had no previous work in this area, I hoped to learn all the phyla in systematic[5] order. However, the only systematic zoology I learned was on my own initiative. Dr. Kepner never advanced beyond hydroids. His interests were cell physiology and the philosophy of science, particularly the concept of vitalism.[6] He belonged to the old school of scientists who were trained in the German universities around the turn of the century. He was not dogmatic and loved a good discussion or logical argument. On exams he would ask questions that implied that animals could look into the future, design their own evolution and make other decisions that could be useful in their evolution. We would write answers that were in conflict with his opinions and philosophy, but if our answers were logical and correct on factual matters he would accept them and give us good grades.

"In graduate school I studied genetics, cytology, and evolution among

---

[5]*systematics: the science of classification; the classification and study of organisms with regard to their natural relationships.*
[6]*vitalism: a doctrine that the processes of life are not explicable by the laws of physics and chemistry alone but are due to a vital principle distinct from these.*

other things, but perhaps the most interesting course for me was one in cytogenetics given by Ladley Husted, student of Professor G. Ledyard Stebbins, University of California. This was at the culmination of the two or three great decades for cytogenetics, and for the first time I became aware of the intricate problems that had fascinated cytogeneticists and of those classic anomalies that persisted to challenge us" (unpublished memoir, 1997).

As one of White's students, Taylor would spend half of each year doing research at Blandy Farm and half at the Miller School taking courses. It was at the Miller School that Herb met his future wife, Shirley Hoover, who

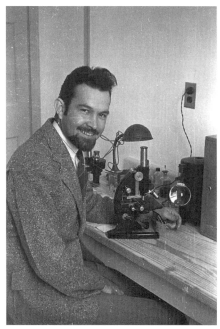

J. Herbert Taylor, at the University of Virginia's Blandy Farm, 1942

Taylor at the University of Virginia's Blandy farm, 1943

Blandy farmers, (left to right) Bernice Speese, Mary Esther Gauldin (Jagger), Earlene Atcheson, J. Herbert Taylor, Summer 1943

was working under Ivy F. Lewis, who headed the Miller School. Lewis was also Dean of the University, an outstanding teacher and grand biologist of the old school. Later Herb recalled:

"I had met her in the fall of 1942 when she enrolled for graduate work at the University of Virginia. At that time we had no dates and the new girl, Shirley Hoover, was more interested in some of the other graduate students. I helped her with some of her classes, especially Dr. Kepner's Zoology course, where she was having some trouble with the lab work. He provided very little guidance for the new students. Shirley came to Blandy for a visit with two or three other students from the graduate school at Virginia in the spring of 1943. In this relaxed setting, I found her a very attractive young lady and ended up spending enough time with her to get know her much better than when we were at the Miller School. She [Shirley] spent the summers at the university's Mountain Lake Biological Station in southwestern Virginia, where her major professor, Dr. Ivy Lewis, would go for part of the summer. I arranged for a visit to Mountain Lake, which I found to be a larger and more formal summer research station than Blandy Farm. Professors from other universities, as well as Virginia, spent summers there teaching.

Shirley C. Hoover, graduate student at the University of Virginia, 1943

Taylor at the University of Virginia, 1943

"However, the war was upon us. As I studied in my room that fateful December 7, I heard by radio of the devastating attack on Pearl Harbor. I then decided to concentrate on the area I knew best and in which I hoped I could finish the research for a doctorate in the time I might have. I wrote a dissertation on the cytotaxonomy and phylogeny[7] of the Oleaceae (olives, ashes, jasmines, lilacs; a family with small chromosomes) under the supervision of Orland E. White before being drafted into service and sent to the South Pacific" (unpublished memoir, 1997).

Only with exceptional help from Dr. White and cooperation of the university was Taylor able to complete requirements for the Ph.D. in two years. A month before he was drafted late in 1943, he and Shirley became engaged. Shirley was still working toward a master's degree.

Taylor at the University of Virginia, 1943

Private Taylor, during basic training period, Spartanburg, South Carolina, 1944

The War Interrupts

For the next two and a half years Herbert Taylor's life was completely different from anything he had prepared for so far; now everything was totally army. He recalled much later how he was soft and hardly ready for the rigors of training that led up to a 25 mile hike with full pack. Despite his lack of readiness, he excelled on the firing range, qualifying as a sharp shooter. In one of those strange

---

[7]*phylogeny: the history or genetic relatedness of a group of species.*

Private Taylor, Spartanburg, South Carolina, 1944

coincidences of war, Taylor's buddy and best friend in basic training happened to be from Roanoke, Shirley's home town. His family were friends with her dad and operated his favorite hardware store.

Taylor's remarkable lucky break came after two months of basic training, when he was transferred out of the infantry and sent to Camp Barkeley, near Abilene, Texas. He was as surprised and flabbergasted as was his commanding officer. Unknown to either of them, a reviewing officer had noticed Taylor because he was a graduate of University of Virginia, the same school the officer's brother currently attended. When he examined Taylor's record he saw the need to reclassify Taylor, according to his skills, and transferred him to the Medical Corps. Life at the next stop, Camp Barkeley, Texas, was easy compared to the earlier Infantry training, and after a short wait in California, Taylor embarked for two years of duty in the South Pacific. So while his buddy remained in the infantry, went to the European theater and was killed in the Battle of the Bulge, Herb Taylor went west and landed in a different theater, the Pacific.

Marking Time in the Pacific

It was through the daily letters Herb and Shirley exchanged throughout those war years that he kept in at least partial touch with his life in science and they developed their life goals. Day by day for the two years, they alternated between onionskin real letters and small, one-page folding "V-mailers" (V for Victory) that had some chance of being delivered more promptly. Army censors seldom made any deletions beyond notifying Taylor never to write on both sides of the paper!

Herb: "The Officers and noncom officers here seem to be a good group, of every kind. A rough bunch, but some very nice fellows from nearly every walk of life. I realize more and more each day how much you mean to me in a thousand ways. You are my contact with the world I've been living in and will come back to" (letter, Camp Croft, South Carolina, 1943).

Shirley: "I know so well how you hate letting your brain mold. But your reading is the only thing to remedy it. Maybe I can find you some perfectly deep and terribly hard book to read. That should keep you in a mental stew! But I don't know of one hard enough!... I will subscribe to *Science* for you, right away" (letter, University of Virginia, 1943).

Herb: "If I can, I'll try to help you in understanding the ways I look at the world.... First I was a dreamer, an idealist. The depression (which left a mark on most of my generation), a hard, practical world, and scientific facts all had a part in making me a realist—a pragmatist in a sense. Now occasionally I drift back to the idealist for a bit of enjoyment, but when I look the world squarely in the face that falls away" (letter, Camp Beale, California, 1944).

Herb: "After 23 days we landed in Noumea, New Caledonia. It was the rainy season and we were housed in tents large enough to stand up inside. The floor was mud and there was a cot but little else. I would enter at night and leave my shoes in the mud while I slid into the sack. The weather was cool and by morning the blankets felt good. During the day the rain fell in showers, and between times the sun shown. It was often steamy and hot during midday. We hiked and drilled some days, but mostly we waited to be assigned. At night we could go to the open air movies. We would wear a poncho and helmet, which would shed the water. If the rain did not fall too fast we could see the screen and hear the sound. I was able to go to the beach and to study the vegetation. Although some of the shore was covered by mangrove swamps with mud and slime, there were also some sandy beaches and we could see the waves breaking on the coral reefs far out from shore" (letter, New Caledonia, 1944).

Shirley: "I now have 30 roaches caught in just half of one day! Dr. Lawson and I operate at 10 a.m. Thursday—with penicillin! Then last night Dr. Cole across the street offered to give me an acetone solution of DDT (crude, that they made in the lab). He

Taylor at his tent, South Pacific General Depot, New Caledonia, 1944

says they really are anxious for information of lethal "dosage"—inaccurate term, and how it affects the insect. How I'd love to be several more persons—or would I? Because I can't do all this at once" (letter, University of Virginia, 1944).

Herb: "Your work with the roaches and penicillin should be quite interesting. I'm glad that you find so many things to investigate—the mark of a good research worker, at least so long as he selects the best and stays with it until something is solved" (letter, New Caledonia, 1944).

Shirley: "I'm not too sleepy to be excited and tremendously interested in the new atomic bomb announced today. It is so amazing—the Flash Gordon stuff—but so frighteningly potent. What a terrific responsibility to guard the use of the thing. We may get into important new things in genetics and cytology but hardly as dangerous, unless how to control sex? Do you suppose the Japanese will change their plans any to escape destruction" (letter, Roanoke, Virginia, 1944).

Not only did Herb encourage and advise Shirley on her research and classwork, he took a challenging break in his routine medical duties to develop and teach a course in physics.

Herb: "My only other accomplishment of note was teaching a physics course in the University of the South Pacific. This came about because the Commanding General of SOPAC decided in the spring of 1945 that the men needed some diversions to occupy their time and prepare them for return to civilian life. The war was winding down and there were many soldiers waiting for rotation back to the states. We would start a university with most of the courses in sciences. In a medical unit there were many biologists, but no one

Sgt. Taylor as Physics Professor demonstrating the principles of simple machines by using pulleys to lift two chairs, University of the South Pacific, Noumea Hdq., 1945

qualified to teach physics. Since I had taught the course in high school, I volunteered. For my efforts, I received a commendation from the commanding general and was featured in a article for the newspaper of the South Pacific!" (letter, New Caledonia, 1945).

Both Herb and Shirley had grown up in rural areas (in Oklahoma, in Virginia) and they always preferred living in the natural world. Even while Taylor served in the Medical Corps in New Caledonia, he found time to explore nature in the tropics.

Herb: "This afternoon I went out in the hills with two fellows from the malaria control office who are looking for mosquito specimens. One is an entomologist, M.S. degree I think, and the other is a zoologist, Ph.D. degree. He is making a collection of bats and rats for the unit and making habitat studies. Anyway I found the most interesting vegetation I've seen on the island—not too far from here either. You see this mountain has the usual Niaouli trees and scrub on the ridges but in the deep ravines down the sides there is a heavy growth with large trees and undergrowth. One of these ravines has a beautiful stream of sparkling water tumbling over the rocks with two waterfalls 20 to 30 ft. high about 150 yards apart.

"Along this stream for a half mile or more as well as down in the valley or canyon below is a subtropical growth that is fascinat-

Taylor with a captured fruit bat, New Caledonia, 1945   *Taylor with* Xanthostemon, *on the river near Pieta, New Caledonia, 1945*

ing—large mango trees, orange and tangerine trees, *Syzygium* and other members of the myrtle family, a tree that superficially resembles our tulip poplar but has a very different nutlike fruit, huge legumes and under that, especially along the edges of the stream are a variety of ferns including the beautiful and stately tree fern, *Cyanthea,* some growing as epiphytes on trees, others as vines with lace-like fronds, another with huge fronds that grow up from the ground (no trunk above the surface). There are several species of palms, a number of shrubs or large herbaceous plants including species of *Solanum,* Verbenaceae, Apocynaceae and Myrtaceae—wild coffee in fruit now, huge banana plants, several species of Araceae (elephant ear)" (letter, New Caledonia, 1945).

Later that year Herb had an unforgettable experience as weekend guest of a native family in a remote area of tropical New Caledonia:

An exerpt from one of Herb's daily letters to Shirley from the South Pacific during WWII, 1945

Herb: "I can say that the three days I spent up near Nakety … were the most perfect I could have spent anywhere, without you. It was the South Seas without soldiers and the signs of war. Some had been there and left their memory and their signs, but all are gone now. If they made a good impression on other families as on the one we stayed with there is a genuine love for the Americans wherever they have been. The little bay with its coconut grove, grass huts and gardens is a land that you dream about but rarely see. It is the South Seas of the storybook with the beautiful maidens taken away and the native music missing. Those things are not missed. In fact the lack makes it more enchanting. The population is sparse and the family is isolated except by water or on foot or horse, no drivable road. But the most remarkable part of it is the family who lives there, particularly the mother. She greeted us as cordially as if we had been old friends and of course we had never seen her before. Our only contact was the letter of introduction by a former soldier guest. Her home consisted of 3 or 4 grass and bamboo huts and a small wooden structure with a tin roof. She took us into one of these huts where there was a small round dining table and a small record player with quite a number of our popular and classic recordings. These, a gift from the soldiers she had befriended, were her prize possessions, her only source of music. She had wine brought out and began to talk with us. Not knowing much English when the first troops came to the island, she has learned to speak fluently and enjoys reading the better American magazines when she can get them. She did this without lessons or a teacher and only a very poor English-French dictionary that she finally obtained. Being completely isolated with only a few magazines the boys send her, she nevertheless has a surprisingly up to date knowledge of world affairs and could talk at our level on almost any subject.

"Madame Galaud's husband works for the nickel company and is the skipper of a fairly large tug. He arranged his work so he could have Saturday off and all of us could go out in the boat to fish, hunt cat's eyes, shells, and have dinner aboard. In the crystal clear water you could see the coral 10 to 15 feet below almost as if there had been no water.

"The contrast between Madam Galaud's attitude, her feelings, to many of her people, makes us feel that our efforts are appreciated, and that after all, maybe these islands are worth saving. Very few, I might say almost none of the colonials arouse any good feeling in us. If I had gone away from this island knowing only Noumea and vicinity I would have no wish to return, but what I saw up there was all an island in the South Pacific should be, not the Hollywood version, but the hunter's, the scientist's, the traveler's paradise" (letter, New Caledonia, 1945).

Herb and Shirley wrote many letters exploring each others experiences, their thoughts, their hopes. And they explored ideas on their future research interests ... how they could work together ... and the practical matter of making a living after the war.

Herb: "At our outdoor movie tonight I saw *The Corn Is Green* and it was good. It reminds me of some scholarships I won that meant the difference between going to school or stopping. It seems now almost a miracle that I got them. Except for the advice and assistance of a high school classmate I might have decided I could not go to college. I had gone to investigate and decided that I could not make it. Then I met Lewis, who was working his way thru [*sic*] college. He showed me a way. It was difficult the first year because I had never really been on my own as many boys are in high school. I had never held a job or bucked the world alone. After that first year I could always take care

(left) Taylor with his buddy, Morris, and little Bernard after their deer hunt near Nakety, New Caledonia, 1945

(top) The Galaud family with two youngest children and a visiting GI, 1945

of myself. The application to graduate school at O.U. … was the only one I made that year, yet it came thru. Then the one I made to Blandy Farm was again the only one I made—rather unusual that I was selected from out there where no one had come to U. Va. before. I had a lot of help along the way, so many to whom I owe thanks for a push or a guide in the right direction and debts I can never repay, at least to them directly. Sometimes I wonder if it was worth their while. Time only will solve that I suppose. There is still a chance that their efforts were not in vain—I hope. I understand how you feel about all those who have a hand in seeing you are free to finish your degree. So you see that we have much to do for those who have helped us as well as for ourselves. Together I'm sure we can do more—maybe make it worth their while" (letter, New Caledonia, 1945).

Herb: "Where do I get my ideas of grandeur? When other boys were playing cowboy and Indian or gangster and their ambition was to be another Buffalo Bill or a railroad engineer or something of the sort, I played—but only halfheartedly. I wanted to be a great artist, author, lecturer or at least a great construction engineer who would build bridges over the largest rivers in far away places" (letter, New Caledonia, 1945).

Herb: "I never found anything in which I could completely lose myself until I started in biological research. That was the last half of my stay at O.U.... Before, research had been a word, a vague concept at best.... By I.Q. tests I'm not a genius but I learn facts pretty fast, and forget them fast. Concepts remain but names slip away. I suppose that is the average person's case. I can analyze material, bring together data and pick the significant points better I believe than the average. Aside from discovery of the new, which is the interesting phase of work, my best bet is in the above field" (letter, New Caledonia, 1945).

Herb: "I know a number of plants that have good chromosomes but whether they have suitable galls or not, I don't know. May apple (*Podophyllum*) has 6–8 very large chromosomes. It is rather easily grown and a very interesting plant which I'm interested in working on sometime. Check on it and see if any gall grows on it.... How about *Peonia?* I think I've seen galls on it and it has 5 large chromosomes" (letter, New Caledonia, 1945).

Shirley: "I spent yesterday afternoon and evening making a card file of all the plant families. I looked up *Podophyllum*. It doesn't seem to get "gally" but some of the Ranunculaceae do, also one species of *Aristolochia*. Peony gets a root gall, hardly what I want. Now Dr. White asks all about my plans for my work and he's made several good suggestions. I do appreciate his interest and I know that it really can mean a lot to me. I reckon he figures he'd better do all he can to see I'm good enough for you!" (letter, University of Virginia, 1945).

Shirley: "Darling, you speak often of being glad that I'm interested in things that you are, that our interests, curiosities, ambitions, and work are all mutual. It is an ideal situation to my way of thinking too.... What I can't wait for is for us to enter our dual research, our dream world thru these past years; it can't fail to be wonderful working together on things we love to do" (letter, University of Virginia, 1945).

Herb: "You paint a lovely picture for us and for my writing, research, etc. Hope I can live up to part of it anyway. You are right, we will go far because we are willing to work for and toward that goal. I feel that with you by my side it won't be hard—at least it will be a pleasant and thrilling journey wherever it leads. Only I'm anxious to be up and doing with you, at the work we love" (letter, New Caledonia, 1945).

Herb: "Now about the sort of job we should look for after the war. That will bear a great deal of discussion as well as investigation. You have as much to say about it as I do and you definitely can be of much help. What we do will depend a great deal on the time and what is available. We can, however, select the region we prefer and try to find something there. I prefer to live in the southern states because so many more things can be grown and I'm interested in improving agriculture and landscaping also for that matter. If we must teach in the beginning, a place near Virginia would be preferred. A place with a state agricultural program for development of pasture grasses, hybrid corn, or some other single crop with little or no teaching is something I would like to try. This might provide a place where both of us could work. Give me your ideas.... If we could be together we could get along much faster with this discussion" (letter, New Caledonia, 1945).

Shirley: "I'm so glad to get your ideas on jobs after the war, both to satisfy my own curiosity and to be able to talk half way sensibly when my friends inquire. I sound rather stupid when I really don't know! Seriously though it sounds so interesting. I liked your preference for the south—I think I agree. And I also think that too often educated southerners leave the south for 'greener pastures' just as they leave the country for the city, thereby impoverishing the former, intellectually speaking, and unfairly so. Certainly the type of agriculture program that you mention would be extremely worthwhile. And I do want to do something that is rather directly important. I imagine it would be too much to hope for that I could also get a job at the same place—wish I could though" (letter, University of Virginia, 1945).

Herb: "I'm much interested in your idea for entertaining students informally. You are quite right about keeping on friendly terms with them, learning to know them. That is one of the traits I've always liked so much about Dr. White. He does it in a different way, but the idea is good. The same applies to the people we work with whether in a university or experiment station" (letter, New Caledonia, 1945).

In October Herb got his first airplane view of the South Pacific when he was transferred briefly to help close out the army's food supply base in New Zealand.

Herb: "The plane ride yesterday was a 'bit tiring' as they say down here, but I enjoyed every mile of the trip. We took off promptly at 7:00 into a partly cloudy sky. Turning south we flew back down the island over the harbor at Noumea. Seeing the island and harbor from the air was a new experience. In fact that shows the real beauty of the island—hundreds of bays, estuaries and mangrove swamps that are beautiful from up there. The numerous small islets that dot the coastline with the coral beneath the shallow water give a striking effect. As we drew away from the island the coral reefs that ring the island could be seen for miles in both directions along the coast. Perhaps two or three miles from the coast is the reef or series of reefs that just reach the water level. The breakers striking against it form a ring of foam and spray that from a mile or so above gives the effect of a chalk line drawn around the island. There are a few breaks in the ring where ships may enter. Inside the reefs the water is always relatively calm. Many of these islands in the South Pacific have the coral reefs but I believe New Caledonia has the most perfect ones.

"Clouds floating beneath obscure the last vision of the place I'd spent the last 16 months. I was glad to be away, though I'll probably go back there in 2 or 3 months, maybe only on my way home or to work a while longer briefly. Some hours later we sighted the extreme northern tip of New Zealand. It is barren for the most part with great stretches of sand dunes, brackish lakes, bays and estuaries with some scrub timber. Shortly we began flying over well ordered farms in the wide flat valley of the sunken rivers. Two station wagons met us and we drove 20 miles south thru a beautiful countryside. All farmhouses are well cared for and attractive. Auckland is a city of 250,000 people, civilization again—streetcars, buses and skyscrapers. What a change!

"I believe we are often inclined to link these islands with Australia, but the flora shows a much closer affinity with that of the islands of the South Pacific. Though I know the genera in some cases, the native plants are quite strange to me except a few which I had seen in New Caledonia or in cultivation. Most of the plants growing around here are introduced species and I found no good book on cultivated plants except ones on English gardens.

"How I wished for color film. The gardens are a blaze of color—green background of well-kept grass and shrubs and beds of Senarerias

[*Cinerarias*], portulacas and pansies—great beds of them of many colors. You would have loved them. I took a picture but the color will be lacking. This is a veritable fairyland now—flowers everywhere and beautiful backgrounds of green—green hills, green forests. Warner was remarking to me yesterday that you hear so much of green tropical islands but none of them compare with this, especially from the air. The green of most of those islands is dark—the jungles or the hills. But here these are bright greens and the flash of brilliant colors. I'd take the subtropical lands every time—I know I'd love Mexico. We must go there sometime, and not too far away either. There are interesting and beautiful parts to the jungle but it gets very monotonous" (letters, New Zealand, 1945).

Herb: "There are many items I want to answer from your letter but first I'll give you the item from Dr. Cross's letter that is of concern to both of us. I had written him a month or so ago about various things but not mentioning postwar plans. Now he says he is interested and has offered me a place at O.U..... Dr. Eigsti has resigned to work for Funk Bros. Seed Co. and his place has been left open. The place has a number of advantages as starting point of our career. The men in the department are young and progressive and both sympathetic toward and interested in research. There would be the added advantage of knowing the university president [Cross] and knowing that he is a research man. It is a fertile field for popularizing and humanizing science. The facilities for research are just fair compared with the large universities, but much better than would be found at most colleges and smaller universities ... [T]hough he did not mention salary the place should carry an

Taylor with two red snappers, Auckland, New Zealand, 1945

assistant professorship with probably $2400 to begin with. It would be difficult to predict the future but if we take it, probably 2 to 4 years would be all we would spend there. As for possibilities for you, it probably offers as much as any place, now that you have your master's degree.… I would like to have your reaction before I send him any word… [T]hinking about the whole thing brings civilian life closer in my mind if not in time" (letter, New Caledonia, 1945).

Herb: "Days follow one another in a meaningless procession. Maybe good for me as time passes and I'm only sweating out the months until this is over and I can see you again and we can take up our work.… When we just do our little tasks without thinking it is not so bad, but if we try to think, to analyze, it is most depressing. I have not answered Dr. Cross's letter and I won't until I hear from you. If the answer is yes and we are lucky enough to get the place we must not allow ourselves to miss all of the travel we have planned. Any that we do will have to come in the first two years probably. There will be some time just after I get back. We can use some of that though we will not want to be moving about all of the time. Even if I should have to teach summer terms there is a month or 6 weeks as a rule that could be used for a trip to Mexico or thru the Northwest perhaps. There is much to see in our own country, even in our own state. Maybe later there will come a chance to go abroad. I'm still hoping your father may be in a position to make some of those trips with us, or us with him. Financially the first years will be the hardest, but we must not let that keep us from going. We must go while we can for in waiting we lose the desire and then the chance. I would like nothing better than to be able to spend most of the summers at the Glen [*Shirley's home near Roanoke*] in later years if we decide to develop it. But there will probably be summer terms if we are teaching, and if not, summer camps and research stations will take up some time for we will want to make some of them for the people we can meet and for the atmosphere as much as anything. There are so many things to do and so little time. I'm most impatient to be about them" (letter, New Caledonia, 1945).

Shirley: "Now to my real talking about Dr. Cross's letter. It was a whopping big surprise to me—how about you? and the more I consider it, the better I like the idea. I vote you accept, if it's what you want, and you think it's the best idea. I've always liked the sound of the place—young men and active department and all that. I see very definite opportunities for our humanizing science (never told you that I used to work summers humanizing religion and made real progress with my mountain folks. Now switching fields shouldn't be difficult at all!) In regard to research facilities, at so many colleges they are just nil—and O.U. would be wonderful in that respect. We would have enough facilities there, and getting immediately into a big university is too much to expect" (letter, University of Virginia, 1945).

(top) Herb relaxes with his younger brother, Tom Dodson Taylor, who has come to visit. They are sitting in the garden of Shirley Hoover's parent's home near Roanoke, Virginia, 1946

(left) Herb Taylor and Shirley Hoover on their wedding day at the Glen, Shirley's parent's home near Roanoke, VA, May 1, 1946

Home at Last ... Science Together

After two years of service in New Caledonia, concluding with a few months in New Zealand, Herb Taylor had been promoted from private to staff sergeant and was ordered back to the U.S. On April 26, 1946, he was discharged in Kansas, and he and Shirley were married in Virginia on May 1.

Herb: "For our honeymoon we circled Florida all the way to the Keys in a used Plymouth sedan. We were lucky enough to meet and talk briefly with the great plant explorer David Fairchild, in the gardens which he had established years earlier, that had been devastated by troops in jungle-training exercises. He was past eighty when we saw him and was teaching his four year old grandson about ants, as they walked to his private gazebo where he was writing another book.

"Shirley and I spent that first year teaching at my alma mater, the University of Oklahoma. We both taught by the Ohio State method, a demonstration, discussion technique that was a new approach for teaching a science to large numbers of undergraduates, and included no lab classes. I also taught genetics and was barely able to keep ahead of my best students. Two events stick in my memory: A fiery redheaded girl in my genetics class challenged me to tell them the chemical and physical properties of the gene. She had searched the library without success, but she was sure that everything was known if we only knew where to look. She screamed at me in class that only my ignorance prevented her from finding the right answer. She was unknowingly partially correct for, unknown to me, O. T. Avery had already published his classic paper on DNA in 1944. However, I am sure that journal was not in our library and only a few people had read and appreciated the significance of his work at that time. Then, at the end of the semester a sassy business administration student who had finished my botany class to fulfill his science requirement told me that I was a good teacher and that he had enjoyed the class, but he advised me that I was wasting my time teaching botany and that I should begin teaching a more important subject in the future" (December Holiday letter 1993).

During that first teaching year Herb and Shirley made a December trip into Mexico in a prewar car, beginning to fulfill their travel dreams of seeing the world. The memorable part of the trip was the adventure of driving back north for a week through Texas in a blinding snowstorm, making barely 100 miles each day, with no windshield defroster and a windshield wiper that stopped working!

Herb had no time for any research that year. Shirley did write her doctoral dissertation, corresponding with her professor in Virginia. She received her Ph.D. degree the following summer.

The next year they moved to the University of Tennessee in Knoxville. Taylor was promoted to Associate Professor and they both taught in the Botany Department. Some time was available for research, and they worked together, beginning to culture a variety of flower anthers to study their meiotic chromosomes. Weekends they often spent hiking in the Smoky Mountains, exploring for big trees and early spring flowers and following rushing streams. Hiking the Smokies Boulevard trail in drizzle to sleep in Mt. LeConte Lodge or climbing up five miles to the flame azaleas on Gregory Bald were equally rewarding. By their third year at the University of Tennessee, they were both teaching full-time and also working every night to finish the interior of their first house. It was a full life—pleasant working situation, good friends, mountain wilderness near, new home in a lovely wooded location overlooking the Tennessee River. They had chosen their road and begun to live their dream.

Shirley Hoover Taylor, Botany Instructor, University of Tennessee, 1948

Shirley and Herb Taylor on a University of Tennessee faculty outing on the Ramsay Cascades Trail in the Smokey Mountains, with the largest yellow tulip poplar tree in the United States, 1949

Chapter 2

The Experiment Heard Round the World

By this time Herbert Taylor had begun to refine his ideas of the direction he wanted to take in his research. It was a time of development of new ideas and new fields within the classic field of genetics. Taylor was ready to be a part of the modern sciences.

Taylor: "I had already made the decision to turn from cytotaxonomy (my Ph.D. project) to the physiology and biochemistry of genetics and cytogenetics. Yet I knew that my training had poorly prepared me for that role. I had read organic chemistry and biochemistry books while overseas, for I had never had a formal course in either. While at the University of Oklahoma, I sat through lectures in biochemistry, and at the University of Tennessee I did the same for physical chemistry.... I met Dr. Franz Schrader at national meetings and had taken to heart his predictions that the cytologists and geneticists would share the labs with chemists and biochemists in the future; and some would become molecular biologists (what Erwin Chargaff of Columbia University, College of Physicians and Surgeons, called biologists practicing biochemistry without a license). Dr. Schrader, who now occupied the chair once held at Columbia University by Thomas Hunt Morgan, had been trained in classical cytology in Germany in the late 1800's, but he saw the changes that were coming before most others. I would have closer associations with Schrader, Chargaff, and Beadle years later, but for now I would remain an obscure loner in my search for pathways in a research program.

"I was not yet [fully] oriented to a new area of research, but I began working with plant organ cultures. My interest was still in cytogenetics, but I wanted to study meiosis in higher plants in what I hoped would be the more maneuverable environment of sterile culture conditions. I had become acquainted with this problem as a graduate student, because Dr. Walton Gregory had begun similar experiments at the Blandy Farm, University of Virginia, where I did the research for my doctoral dissertation.

"Besides the culture of plant tissues I began learning to use radioisotopes to trace the metabolism [of nucleic acids] in my cultures or even in intact plants. My lab was only 20 miles from the rapidly developing Oak Ridge National Laboratory that grew out of the research and production facilities that had been used during the war to produce the atomic bomb. Soon after I came to Tennessee, I went out to the National Lab to see the newly appointed Director of the Biology Division, Dr. Alexander Hollaender. He was just beginning to recruit the scientists that would make the laboratory one of the major research centers in the country in a few years. He was not particularly interested in my research, since the objective of the Lab was to study the effects of radiation on living systems. This was a part of the program of the parent organization, the Atomic Energy Commission, whose interest was to develop nuclear energy for power and various uses in medicine and other civilian uses. The biological effects of the various types of radiation were a major concern. The production of nuclear weapons had been shifted to other sites, but security was still a major concern and one had to have FBI clearance and enter the lab via guarded gates for many years.

"I was interested in being a consultant at the lab because it would give me access to lab facilities where I could use radioisotopes. I could not handle radioisotopes in our limited and crowded lab space at the University of Tennessee. A group of us presented preliminary statements to a recruiter for the Naval research fund. My project was to study cellular differentiation by using the plant pollen grain, which begins as a spore. The Navy representative picked that kind of problem above all others presented, as an example of the kind of fundamental research that the Navy might be willing to support. The audience was stunned by the outcome, but Dr. Hollaender made me a consultant even though he did not think I had very interesting problems and [believed I] lacked promising techniques to study them.

"[At Oak Ridge] I began working in the lab of Ray Noggle, who had once worked at the Blandy Farm, where I did the research for my Ph.D dissertation. Within a short time I was working with radioisotopes, which were just being introduced for routine use. Phosphorus-32, sulfur-35, and carbon-14 were available, but few organic molecules had been produced, and the methods for tracer work were just emerging. I began studying the incorporation of phosphorus-32 (^{32}P) into the DNA of cells in the anther. I used the buds of Easter lily as the source for my anthers. It is not only a large bud, but Ralph Erickson had recently shown that the stage of meiosis could be predicted by carefully measuring the length of the bud. This allowed dissections at predictable stages either before or after allowing incorporation of a radioisotope, atoms that decay with a predictable half-life and reveal their presence by the production of beta particles, which can be detected by a Geiger counter or photographic emulsion. I was soon making microscopic preparations containing ^{32}P and detecting the isotope by a thin film of photographic emulsion. When the film is placed against the specimen for an appropriate time and then developed in a dark room like a photograph, the film will contain silver grains, dark spots visible with a high power microscope. If the specimen can be viewed in contact with the film the location of very small amounts of ^{32}P can be determined. The combination is called an autoradiograph, and this device became a very important tool in my research for years to come. I soon learned that the weaker the beta particle (the lower the energy) emitted by the isotope, the higher the precision with which I could locate the isotope."

These were Taylor's first trials with a technique that, when perfected, would revolutionize the field of genetics and cell biology: the first experiments that would allow DNA to be traced over time. Later these techniques would be used by thousands of scientists to study the growth of cells in studies as diverse as genetics, cell biology, cancer biology, and immunology.

But much more work was needed to perfect the technique....

Taylor: "I had tried to prepare autoradiographs to locate the ^{32}P in cells from my cultured anthers, but I had little success. I had microscope slides coated with photographic emulsions, but they were damaged by mounting specimens on them. About that time I got a break. Dr. Alma Howard from England visited Oak Ridge, but because she had failed to get FBI clearance, she could not enter the lab. A small group of us met her at her motel room, where she described some recent experiments using ^{32}P and autoradiographs to study the incorporation of the isotope into the DNA of bean root cells over the cell cycle. I was intrigued by the results, but much more importantly I realized immediately that her materials and techniques for making autoradiographs were far superior to those I had tried to use.

"While my enthusiasm was high, I went to Dr. Hollaender and asked him to give me a small lab where I could use isotopes and autoradiography to study the synthesis of DNA in the meiotic cycle and the development of anthers. He was less impressed than I and declined to support my project. I soon began casting about for other positions where I could have more space and facilities for research using radioisotopes. Within a few months I had firm offers from two universities. One was the University of North Carolina in Chapel Hill and the other was in the Department of Botany at Columbia University in New York City, where I could occupy a suite of rooms recently vacated by Marcus Rhodes, who had moved to the University of Illinois. I had followed and admired his work on corn genetics since I was a graduate student and would be proud to follow in his footsteps. On the other hand we did not like large cities, and I would take a cut in pay and drop back to assistant professor level. I knew North Carolina was one of the best universities in the South, their Department of Botany had some good people, living would be easier. The decision was a tough one, but after Shirley and I considered the options, we decided to be brave and try the Big Apple. It was one of those major turning points, and I will never know where the other road would have led, but I am reasonably sure we made the correct choice. So we left our new house when school was out in 1951 and headed for New York" (unpublished memoir, 1997).

With their new baby girl the Taylors arrived in Manhattan at the time when signs at every exit from the city read "This road will be closed to all traffic in case of nuclear attack." They moved into an 8th floor apartment with a history. This 10-story Columbia building had been constructed about 1900 and was among the first to be wired by Edison's initial plan for dis-

tributing electricity—direct current. As a result, many of their electric items (refrigerator, iron, clock, record player) were unusable or required insertion of a current converter in the line. Faced with so much change, and finding foods available only in separate specialty shops rather than supermarkets, they got the feeling they had moved halfway to Europe, but they were directly across the street from their university lab in Schermerhorn Building. In 1997 Taylor recalled those years at Columbia and summers at Brookhaven Laboratory on Long Island.

◇

Building His Breakthrough Experiment

Taylor: "That first year Shirley assisted me in the lab. We could leave our baby sleeping in one of the lab rooms.... I had a light teaching load and could spend most of my time preparing for my research, which I hoped to get under way soon. I had plenty of room but very little equipment or supplies. The department gave me very limited funds, and I applied to the Atomic Energy Commission for a grant, which was funded with a few thousand dollars.... I was fortunate in that I arrived on the scene when research grants from government agencies were being expanded, and because my research was successful in discovering

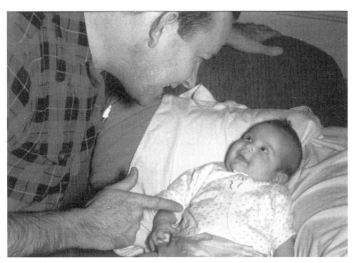

Herb talking to his daughter, Lucy, in Manhattan, New York, 1957

new and interesting aspects of genetics, I maintained support from the same agency for my entire career.

"I also visited the Biology Department at the Brookhaven National Laboratory and applied to be a research collaborator. This was what they called a scientist visiting from neighboring universities.... Beginning in the summer of 1953, we spent the summer months living in rented housing operated by the laboratory or in small rental houses near the coast. Brookhaven is located inland near the center of Long Island. At that time the area was not heavily populated, and there were orchards and vegetable growing farms nearby. It was a pleasant place to spend the summers, and the laboratory was a desirable place to work. Many facilities were available, and the permanent and visiting scientists were helpful and stimulating colleagues. I worked first in the lab of Monty Moses, who had recently obtained his doctorate in the Zoology Department with Professor Arthur Pollister, who had been at Columbia for many years when I arrived. His field was cytology, and I found more in common with him and Professor Frans Schrader than with the other botanists in the department" (unpublished memoir, 1997).

"During the first three or four years my research revolved around the problems of meiosis in plants, the two cell divisions in which chromosomes pair and exchange segments in the formation of four haploid spores (haploid describes those cells which contain a single set of chromosomes). Mitosis is a relatively rapid process that occurs every 12 to 24 hours, but meiosis is slow. In lilies it takes one to two weeks, and in frogs and salamanders it can last for weeks or months. The time of synthesis of DNA (chromosome reproduction) had been determined for the mitotic cycle but was unknown for meiosis. I spent those first years trying to determine when the DNA was replicated during the long meiotic cycle in several different species. The underlying problem was to learn more about the mysterious process of crossing over, exchange of segments between the set of chromosomes derived from the male parent and the set from the mother (female parent). One unsolved question was whether DNA replication and chromosome reproduction was part of one series of events. We knew that the chromosomes contained other components such as several kinds of proteins and RNAs. The role of these non-DNA components in structure and reproduction was unknown.

"To study these problems I teamed up with Monty Moses at Brookhaven National Laboratory my first summer. He was using microspectrophotometry to study individual cells, and I could use radioactive molecules and autoradiographs to supplement the studies. As soon as I had enough grant funds I employed one of Arthur Pollister's recent graduates, who also knew the microspectrophotometric techniques, to help me at my Columbia University laboratory. We published several papers, but they received relatively little attention from the scientific community. Nevertheless, I was establishing a position by extensive studies" (unpublished memoir, 1997).

Despite Taylor's modesty, other people were noticing. In The Cells of the Body: A History of Somatic Cell Genetics *(Cold Spring Harbor Laboratory Press, 1995), Henry Harris describes the Taylors' work from this period, "[Earlier work] was expanded by the experiments of Taylor (1953), who studied the incorporation of radioactive phosphorus into the chromosomes of cells undergoing mitosis and meiosis in* Tradescantia *and* Lilium. *Taylor and Taylor (1953), working with* Tradescantia *and* Lilium, *simultaneously administered* ^{35}S *to label proteins and* ^{32}P *to label nucleic acids, and they were able to discriminate between the two labels by treating the preparations with trypsin before submitting them to autoradiography. Whereas protein synthesis was found to take place continuously, DNA synthesis in cells undergoing either mitosis or meiosis was again shown to be limited to one stage in the cell cycle." Indeed, he was establishing a position in cytogenetics.*

In the Columbia area of Manhattan, the Taylors found stimulating opportunities to broaden their horizons. Outstanding lectures at Union Theological Seminary, Pete Seeger concerts, League of Women Voters friends at Columbia Teachers College, and discussion groups at Riverside Church in the tradition of Harry Emerson Fosdick's "social gospel" were all stimulating. At Riverside Church they found for Lynne a nursery school founded by Sophia Fahs (outstanding humanist educator), but in contrast to their life at the University of Tennessee, there was no wild country to explore, only manicured parks to visit and car trips to see fall leaf colors. Summers at Brookhaven Lab were their only relief from regimented city life, but at Columbia, genetic research opportunities were good. The field of genetics was coming to a major turning point.

His first graduate student, Sandhya Mitra, remembers Taylor's excited *discussion with his class about the newly published paper by Watson and Crick describing a theoretical structure of DNA. And how Taylor immediately began thinking about how one could demonstrate it in chromosomes—prove that the predictions made by the model for DNA would hold up.*

Mitra: "One day in 1953, Taylor wrote the words DNA DOUBLE HELIX on the classroom board. He mentioned that a momentous report had come out in *Nature*. This announcement might well be the harbinger of a sea change in the understanding not only of the process of inheritance but also of the entire field of life science. I remember the inspired look on Herb's face and the sensation of gooseflesh while listening to him analyze the historic model of Watson and Crick. The entire class was enthralled by the prophetic way in which he speculated about the significance of the double helix model for a plethora of novel investigations in biology. Taylor immediately devised strategies to corroborate the validity of the model. Taylor kept bean stems in a beaker under the light of his desk lamp. Later, squashes were made of bean tissue and the slides stained and autoradiographed in due course. As Taylor's research assistant, I was entrusted with much of the dog work, not quite knowing the goal of this particular series of experiments. I later realized that this was the pilot study for the investigations that were eventually carried out at the Brookhaven Laboratories and that proved the semiconservative nature of replication of the DNA molecule" (personal letter, 2000).

Later Taylor himself wrote about how he came to that point in 1954 with all the technical tools ready to do the experiment that he did not yet know awaited him, as if all he had done before was preparation for the experiment to come.

Taylor: "In 1954 I began writing a chapter on 'autoradiography at the cellular level' for a comprehensive three-volume work on physical techniques in biological research edited by Gerald Oster and my colleague Arthur Pollister. In the process of searching the literature, I found that the radioactive isotope of hydrogen, tritium, would give the highest resolution in autoradiographs because the beta particle emitted upon decay has a very low energy. It would move on the average less than a micron in a photographic emulsion. Other isotopes useful in biological research such as phosphorus-32, carbon-14, and sulphur-35 were impossible to locate with precision because their beta particles had enough energy to move a few too many microns in such emulsions. However, the hydrogen isotope had been so little used that I hardly mentioned it in the long review. It would soon become available to researchers, because it was produced in the process of making hydrogen bombs.... The most important event of the period, however, was Watson and Crick's model of DNA and their proposal for a mode of replication to pass on its specific base sequence. Dr. Morris Freidkin had prepared ^{14}C-labeled thymidine and shown that it was used exclusively in the synthesis of DNA and could be used for autoradiography and the study of DNA synthesis at the cellular level. Dr. Daniel Mazia and his postdoctoral fellow, Dr. Walter Plant, had obtained some of the ^{14}C-labeled thymidine and studied its distribution in cell division in the root tip cells of *Crepis capallaris,* which has only six chromosomes. Resolution with ^{14}C was not good enough to reveal how chromosomes segregated, but they concluded on the basis of their experiments that the double helix was separating as a unit—that segregation of DNA was not semiconservative as proposed in the Watson-Crick model" (unpublished memoir, 1997).

As Taylor tells the story of planning that historic experiment, one feels drawn into the moment—the beauty, the simplicity, the elegance of the experiment! How brilliant to realize that this was the experiment that would demonstrate how DNA worked and that he, perhaps alone, was in a position to do it!

Taylor: "I realized that the only way to answer the question was to label thymidine with tritium (^{3}H); that would allow me to follow the distribution of tritium in individual chromosomes. Very few attempts had been made to label nucleosides with tritium, and exchange from tritiated water gave very low specific activities, not useful for the autoradiography I wanted to do [where I realized I would need high specific activities].

"Part of the next year was spent in learning all I could about tritium and how it might be used to label DNA in chromosomes. Very little was known, and the scientists I approached who had worked with tritium showed little interest in helping me, probably because they could not see that tritium would be very useful in biological research. The techniques available for detecting and counting tritium with its very low energy beta particle were few. Carbon-14 was the isotope of choice for most of the work in biochemistry. I learned that carbon-14 had been used to label thymidine, a water soluble substance that was used by cells almost exclusively to synthesize one of the four nucleotides of DNA. If roots were immersed in a solution of thymidine, the nuclei of cells that were about to divide would be labeled while other parts of the cell and nondividing cells remained free of the bound thymidine. If I could learn how to label thymidine with tritium at a high specific activity, that is, to have a high ratio of the molecules in a solution contain one or more tritium atoms, I would have the best possible tracer for DNA in chromosomes in studies using autoradiography. The problem was that I was not an organic chemist and knew little about the synthesis of organic molecules. Those whom I contacted were not interested in spending their valuable time in making a compound that they did not require for their work.

"After two summers working in the [Brookhaven] lab with Montrose (Monty) Moses, I teamed up with Philip Woods, whom I had known when he spent two years as a postdoctoral fellow before taking a position at Brookhaven. We [planned to] try to label thymidine with tritium [^{3}H] and use that radioactive substance to selectively label the DNA in chromosomes. After I arrived at Brookhaven and we were in the planning stages for our experiments, we learned from colleagues that Walter L. (Pete) Hughes also wanted to label thymidine at a high specific activity and use it to kill cancer cells. He had a plan for the labeling that promised to work, and after we explained our experiments, he agreed to supply us with some of his thymidine when and if his experimental preparation worked.

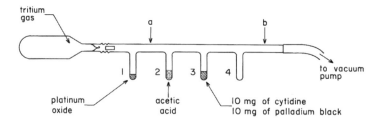

Diagram of apparatus and technique used by Taylor, Woods and Hughes to produce the first tritium labeled thymidine.

"My plan was to grow seedlings of broad bean, *Vicia faba,* in a solution with ^{3}H thymidine for eight hours and then rinse and transfer the roots of the seedlings to a new solution free of radioactive material. I knew from some experiments that Alma Howard and S. R. Pelc had recently published using phosphorus-32 to label DNA in these roots, that the cell division cycle required about 24 hours. After a division the root cells in the growing region near the tips would go through a preparation period of about eight hours before beginning to synthesize their DNA in preparation for another cell division. The period of synthesis would be complete in about eight hours, and the cell would spend another eight hours before dividing, a process that required an hour or less.

"Within a few weeks Hughes brought me his first batch of ^{3}H thymidine, and I carried out my first set of experiments. I did not know how long I would have to expose the autoradiographs to accumulate enough grains to detect labeled chromosomes, but I developed some of the autoradiographs in two weeks and reserved the remainder of the slides for later development. I was disappointed to find no labeled cells. However, Pete Hughes told me that he had prepared a second batch of thymidine of higher specific activity. I repeated the experiments and stored the autoradiographs in the refrigerator to expose for about two or three weeks. When these were developed I saw that some chromosomes and many interphase nuclei were labeled so that grains appeared in the photographic emulsion that remained attached to the glass slides on which the cells were squashed. The resolution of the site of decay of tritium was as good or better than I had expected" (unpublished memoir, 1997).

The preliminary experiment had worked! Taylor knew now that he could use this method. The tritiated thymidine was being incorporated into the nucleus and specifically into chromosomes, and furthermore the resolution was better than anything ever used before. No one had ever been able to put all these steps together before, and this was the technique that could answer the biggest questions in the field of genetics of the times—was the theoretical structure for DNA as proposed by Watson and Crick correct? Was DNA a double strand, complementary and anti-parallel? And was that the structure found in the chromosomes of eukaryotes?

Discovering the Principle of Semi-Conservative Replication

Now Taylor had demonstrated the ability of his technique to label chromosomes and that the resolution of the site of decay was good enough to distinguish between chromosomes. He was ready for the final experiment.

Taylor: "The second part of the experiment was crucial. We grew cells in labeled thymidine (in roots) for several hours and then transferred them to a solution containing colchicine. We knew this drug would block the separation of chromosomes, and retain them in the same cell, but allow them to proceed through another replication cycle after we removed the thymidine" (unpublished memoir, 1997).

Taylor: "Chromosomes in cells that reached division within 8 hours after removal from thymidine had both chromatids labeled. However, after a longer time, when I found cells with two times the normal number of chromosomes, I saw that one chromatid of each chromosome was labeled while the sister chromatid was free of labeled DNA. **It soon became clear that chromosomes must consist of two subunits, which are separated during replication. A labeled subunit is synthesized along each, and at the following division both new chromosomes**, which often remain attached at the centromere in colchicine treated cells, [only] *appeared* **to be labeled** because the subunits are too close together to be resolved. However, the two subunits remain intact and are segregated at the next replication. At this second replication only unlabeled DNA is synthesized along each subunit. Therefore, **only *one* of the two chromatids** (new chromosomes) has a labeled subunit made during the previous replication in tritium-labeled thymidine. It registers on the photographic emulsion, but the sister chromatid does not. **I was both pleased and surprised that the results were so clear and followed the prediction of semiconservative replication of DNA.** Of course, I was aware that the chromosome and the DNA double helix were orders of magnitude different in size. Why a whole chromosome would segregate as predicted for the DNA helix would puzzle us for years to come.

"I showed the cells to some of my colleagues in the Biology Department, but soon many others came to see the results and the remainder

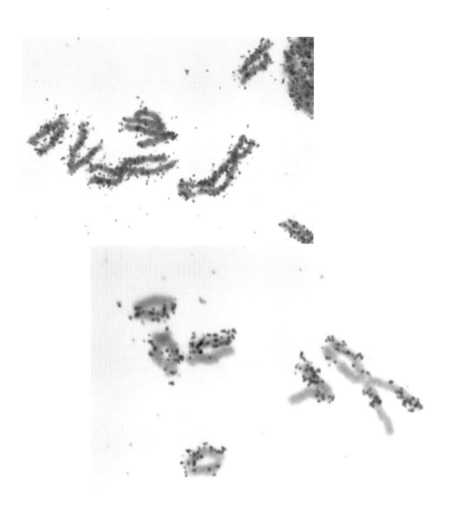

Chromosomes of Vicia [faba] *[top photograph] at the first division after incorporating tritiated thymidine during DNA replication. All chromatids are labeled [bottom photograph] at the second division following incorporation of tritiated thymidine during one replication cycle. The segregation of labeled and unlabeled chromatids is complete and regular except that sister chromatid exchanges occur during the replication when labeling occurs and at the second replication. The frequency of exchanges is relatively high in this cell since colchicine was applied to the roots after the second replication of this complement of chromosomes* (Int. Rev. Cyt., *1962).*

of that day and the next was largely devoted to show and tell sessions. We soon noted that chromatids in many instances were labeled for only part of the length, but in each of these instances the sister chromatid was labeled in the complementary segment. It appeared that a reciprocal exchange (sister chromatid exchange) had occurred at some time during or following replication. In the *Vicia* root cells, the exchanges were frequent enough to be annoying when one wished to demonstrate regular semiconservative distribution of DNA. **However, I was pleased that I could detect the exchanges, for one of my objectives in these experiments was to see if I could use autoradiography to study physical exchanges correlated with genetic crossing over during meiosis**" (*Trends in Biochemical Sciences,* 1997).

The News Breaks Around the World

Taylor's proof of semiconservative replication was the first proof of the **Watson-Crick** *structure for DNA in eukaryotic chromosomes.*

Taylor: "There was so much interest in my demonstration of semiconservative distribution of DNA at the chromosomal level that I realized I would have to defer studies of crossing over for a while. The summer period for my work at Brookhaven was drawing to a close and I would soon return to my home base at Columbia University for the fall term. In the meantime I attended the AIBS meeting in Storrs, Connecticut, where I was to present a short paper on a different topic for which I had submitted an abstract. The chairman of the session asked me to present instead our recent work on DNA and he agreed to extend the time. It was a warm September afternoon and because rumors had spread about our experiments, the room was overflowing and many who could not get in listened at the open windows. A few days later an international meeting on DNA was held in Tokyo, which I did not attend. However, several people from Brookhaven and some from the Storrs meeting attended and carried word of our experiment to Tokyo and soon our results were relayed around the world. A few weeks later, after I returned to Columbia, Francis Crick called to ask if I would supply a photograph for him to use in an article he was writing for *Scientific American.* In the meantime, at Columbia, my friend and mentor, Professor Franz Schrader, was alarmed that my experiments were known around the world, but no publication or even an abstract had appeared in print. Even though I was convinced that our experimental results were clear and correct, the photographic evidence was not as good as I wanted. Nevertheless, I put together what was available and Professor Schrader communicated it to the *Proceedings of the National Academy of Sciences* for quick publication" (*Trends in Biochemical Sciences,* 1997).

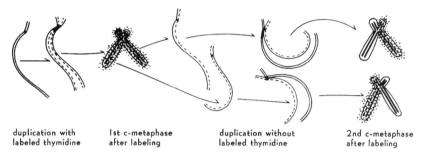

Diagrammatic representation of the organization and mode of duplication as revealed by autoradiographs. The two strands necessary to explain the results are shown, although these are not resolved by microscopic examination. Solid lines represent non-labeled strands, and dashed lines represent labeled strands. The dots represent grains in the autoradiographs (Proc. Natl. Acad. Sci. USA, *Fig 3, 1957).*

Taylor: "I immediately began experiments with another plant I had available, *Bellevalia romona* of the lily family. It has only four pairs of large chromosomes, three of which are easily distinguishable by their morphology. One problem was that I had to grow roots from bulbs and I had only a few bulbs, each of which produced only a few roots. In spite of these technical difficulties, I obtained autoradiographs superior to those of *Vicia* and there appeared to be fewer sister chromatid exchanges in *Bellevalia* chromosomes.

"The fall term was a busy one with visitors often coming by the lab. The most impressive group was Francis Crick, Max Delbruck and Alexander Rich who arrived unannounced and asked, no demanded, to see the evidence for semiconservative replication. Fortunately, I had a cell of a *Bellevalia* root at the second division after the labeling cycle mounted under my microscope. Since I now knew the segregation did not occur until the second division, I had delayed the colchicine treatment for a longer time after labeling so that I had a diploid cell with only eight chromosomes. The segregation was clear and there were only a few sister chromatid exchanges. I believe my visitors went away convinced" (*Trends in Biochemical Sciences,* 1997).

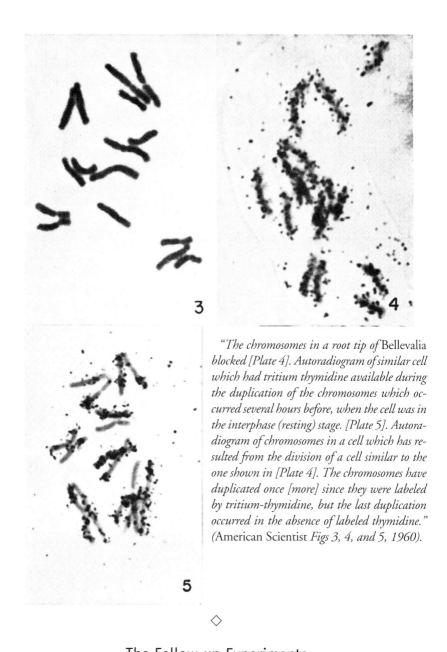

"*The chromosomes in a root tip of* Bellevalia *blocked [Plate 4]. Autoradiogram of similar cell which had tritium thymidine available during the duplication of the chromosomes which occurred several hours before, when the cell was in the interphase (resting) stage. [Plate 5]. Autoradiogram of chromosomes in a cell which has resulted from the division of a cell similar to the one shown in [Plate 4]. The chromosomes have duplicated once [more] since they were labeled by tritium-thymidine, but the last duplication occurred in the absence of labeled thymidine.*" (American Scientist *Figs 3, 4, and 5, 1960*).

◇

The Follow-up Experiments
Sister Chromatid Exchanges demonstrated polarity of the DNA Double Helix.

Taylor: "Unknown to me at the time, in my audience at Storrs was A. H. Sturtevant, who was very interested in the results and possibilities of its future use for studies of crossing over. I knew him only casually at that time, but his photograph along with that of T. H. Morgan hung in our seminar room at Columbia because the famous fly room, where his early, classic *Drosophila* experiments had been performed, was just down the hall. Morgan and Sturtevant had moved to Cal Tech many years before I came to Columbia. Professor Morgan was no longer living and Sturtevant was retired, but he still came to the lab regularly as I learned later when I spent my sabbatical leave at Cal Tech. Before long I had an invitation from George Beadle, now Director of the Biology Division, to come to Caltech and give a seminar on our work. I found the group very friendly and informal; George Beadle operated the slide projector for me and Linus Pauling came over from Chemistry and sat on the front row. The next morning Max Delbruck called one of his work sessions to explore the possibilities of learning more about the nature of the chromosomal subunits containing the DNA by analyzing the pattern of sister chromatid exchanges. He pointed out that if the units were unlike (had different polarity, for example, like the two chains of the DNA helix) the exchanges would occur in pairs. At the second division after labeling, both chromosomes descended from an original labeled chromosome would each have an exchange at the same locus; e.g., the exchanges would occur as twins. He also admitted that his recently published hypothesis that exchanges might occur between each turn of the helix to resolve the problem of unwinding the DNA helix in replication was ruled out by my experiments. Such exchanges would randomize the label unless it was assumed that the exchanges occurred at precisely the same site at each replication, an unlikely possibility.

"Back at my lab, I considered the difference between the frequency of twins expected if the subunits were unlike and the frequency if they were alike and free to rejoin at random. It turned out Delbruck had neglected the exchanges that might occur at the second replication. These exchanges would be singles whether a polarity existed or not. A calculation showed that if one analyzed the tetraploid cells at the second division, the frequency of twins for units having polarity would be one twin from the first replication to two singles at the second division because there would be two times as many chromosomes at the second replication. However without polarity the ratio would be one twin to ten singles. With these large chromosomes and only 16 in each tetraploid complement, the difference between the two hypotheses should be easy to determine.

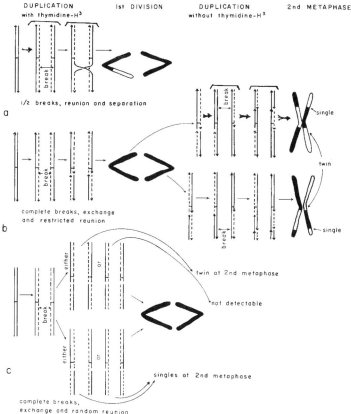

*Chart showing the results of chromatid exchanges and the **predicted ratio of twin and single exchanges.** a) The predicted results of one-half chromatid exchanges, which were not observed. b). The results of exchanges when reunion is restricted by a difference between the two strands of the chromatids. c). The results of exchanges when reunion is unrestricted and rejoining of strands occurs at random. Labeled units are shown as dashed lines. The difference between strands is represented as a directional sense and is indicated by arrows (Genetics, Fig. 9, 1958).*

"I soon analyzed enough to be sure the ratio was far from 10 singles to one twin. In fact the twins appeared to be too frequent to fit either hypothesis unless more exchanges occurred at the first replication than at the second. The ratio was 81 twins to only 30 singles. That result was included in a paper submitted to *Genetics* (Taylor, 1958), but by the time it appeared or soon thereafter I had additional results, which approached the predicted 2:1 ratio. That result was included in a review given in 1958 at the International Congress of Genetics, Montreal (Taylor, 1959)" (*Trends in Biochemical Sciences*, 1997).

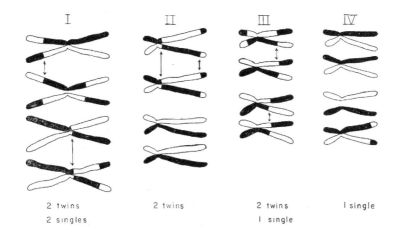

*Diagramatic reconstruction of the **chromosome complement** from a tetraploid c-metaphase at the second division after **labeling to show the position of exchanges.** [Singles are events occurring in only one of the two chromosomes of the pair; twins are events which occur at the same point in both members of the pair]; twins indicated by arrows. (Genetics, Fig. 8, 1958)*

By 1959, Herb Taylor had published evidence not only that the DNA of eukaryotic chromosomes replicated semiconservatively but also that the strands of the double helix of DNA in chromosomes had opposite polarity. Taylor's diagrams demonstrated clearly what could be expected if DNA was parallel or anti-parallel (see reprinted diagram above).

Taylor: "What would allow unwinding and segregation as demonstrated was a problem that led to various **models for chromosome structure.** One early model was dubbed 'the centipede' with DNA attached to a central axis. This was soon abandoned for a linear duplex with hypothetical interruptions in the DNA at fixed intervals and folding to account for the differences in size of a molecule compared to a chromosome. After Messelson and Stahl (1958) demonstrated semiconservative replication of bacterial DNA at the molecular level and later when Simon (1961) demonstrated similar replication in mammalian cells the case for a single DNA helix running the length of a chromatid became much stronger. However, it was only after topoisomerases[8] were demonstrated and their action understood that one could better under-

---

[8]*topoisomerase: an enzyme that catalyzes the relaxation of supercoiled DNA.*

stand the possibility of folding a single helix up to a meter long into a chromosome, and yet have regular segregation of subunits and only a relatively few sister chromatid exchanges" (Taylor, 1997).

Many were excited by the new data that Taylor had provided, and his first model gave early molecular geneticists a way to start thinking about fitting the data to the concept of the chromosome. Max Delbruck was one of many who were enthusiastic. The data were simple; the logic was brilliant; the consequences for understanding DNA and the chromosome were astounding.

Delbruck: "Nobody could be more amazed and excited than I by your wonderful finding. It is so rare in biology that a simple logical argument points the way to a discovery and that an experimental procedure is so powerful that it can right away give an unambiguous answer" (letter to Taylor, February 1958, from *Meselson, Stahl and The Replication of DNA,* Frederic Holmes, Yale Univ Press, 2001).

At the Same Time

Taylor: "One unsolved question was whether DNA replication and chromosome reproduction were part of one series of events. We knew that the chromosomes contained other components [in addition to DNA] such as several kinds of proteins and RNA's. The role of these non-DNA components in structure and reproduction was unknown" (unpublished manuscript, 1997).

Taylor hoped that, as a companion to the experiments with tritiated thymidine, he could make tritiated precursors to RNA and to proteins so that he could track their involvement in chromosomes as well. He worked in the lab of Pete Hughes preparing tritium-labeled cytidine and arginine. None of these labeled compounds was yet available commercially. Cytidine was to be used to study RNA synthesis and arginine to determine whether it could be demonstrated how the chromosomal proteins segregated during chromosome reproduction. The protein demonstration was unsuccessful, but labeling of RNA proved interesting, and Herb Taylor and Phil Woods spent the next months studying RNA metabolism by autoradiography.

Just as when DNA was labeled, the label of tritiated cytidine appeared immediately dispersed throughout the nucleus if the cell had been in interphase, but unlike the labeled thymidine, labeled cytidine disappeared from the nucleus within a short time after labeling and reappeared in the cytoplasm. ***And the labeled cytidine never showed up in mitotic chromosomes, proving that RNA was not a component of chromosomes.*** *Taylor continued to wonder whether RNA was double stranded and what its template might be! What he had just demonstrated without realizing it, and what no one else appreciated at the time either, was that RNA was made in the nucleus during interphase and rapidly exported to the cytoplasm.*

Jesse Sisken, Taylor's graduate student in this period:

Sisken: "… not only were Taylor's experiments the beginning of the use of all tritiated compounds in biomedical research, but that one compound alone, tritiated thymidine, opened doors to the study of cell proliferation that had not even been imagined then. It is hard to imagine what research in cancer or immunology, for instance, would be like today without Herb's pioneering development more than 40 years ago. The imprint of this modest man's work has been and continues to be enormous."

Chapter 3

Fame Abroad

After he gave one of the principal invited papers at the International Congress of Genetics in Montreal (1958), Taylor flew to Paris and on to Geneva to spend a month demonstrating his technique of and results from the autoradiography of chromosomes at the 2nd Conference on the Peaceful Uses of Atomic Energy. Alexander Hollaender was in attendance. Hollaender had been the new Director of Biology Division at Oak Ridge years earlier when Taylor had approached him with his ideas of starting work with ^{32}P. Then he had been a rather lukewarm supporter; Hollaender was now in strong support of his research! Taylor wrote home to Shirley:

Taylor: "Last night Alexander Hollaender invited me to a round table dinner, from which we went to meet the Russian geneticists who are here—anti-Lysenko.[9] The group we met in Montreal last week was only of the pro-Lysenko group. We got a good picture of what is going on, I think. They are being allowed to set up institutes and work on genetics with classical methods, but they feel strongly that they must produce

---

[9]*Lysenko: Trophim Denisovich Lysenko (1898–1976), a Russian geneticist who dominated the field in his country from the 1930's to the 1960's and perhaps beyond. He believed that individuals could acquire traits and pass them to the next generation. The classic example was giraffes—Lysenko proposed that members of each generation acquired the trait of long necks by stretching their necks to reach leaves higher on the trees, and their offspring could inherit that acquired trait. His views on the acquisition of traits were supported by the state.*

practical results in all their work, especially in the beginning. This is the most potent argument that Lysenko uses against them....

"Our work and the other work on DNA has been one basis for the comeback of classical genetics. [Russian] physicists and biochemists have become aware of the work, and besides being intrigued, they see that there is a genetic substance in the chromosomes that has continuity from cell to cell and generation to generation. This gives them a basis for rejecting the ephemeral ideas of the Lysenko group.... They are very interested in the theoretical problems, however, and I find that I am famous in Russia. They were impressed, and a little disappointed maybe, that I was so young. They had pictured Taylor as one with a gray head I think" (letter to Shirley, Geneva, September 1958).

Suburban Living and a California Sabbatical

After six years of apartment living in Manhattan, the Taylors had bought a house in suburban Westchester County that gave them some freedom to garden and their three children the chance to play outdoors. To broaden their children's education they located a Friend's Preschool and First Day School in nearby Scarsdale. Many of the members of this Silent Quaker Meeting were United Nations staff and from various countries. There too, the Sophia Fahs comparative religion materials were the teaching core, and Shirley taught one of the weekly classes. Following the move to the suburbs and with three children to care for, she was now full-time family manager, and her involvement in science was limited to some writing and editing with Herb.

Taylor children (left to right) Lucy, Michael, and Lynne, Pasadena, CA, 1958.

In the winter of 1958, taking their two little girls and six-month-old son along, they drove west to Pasadena, where Taylor had a Guggenheim Fellowship to work in Professor Max Delbruck's lab at Cal Tech. Six months in southern California was a pleasant interlude for the whole family.

Taylor: "The work moved slowly on my problem of crossing-over in grasshopper spermatocytes using tritium-labeled thymidine and autoradiography. I was fortunate to have Matthew Meselson down the hall, and I was able to observe first hand the technique and results of his use of the ultracentrifuge and a density label to demonstrate the semiconservative replication of DNA. Just below was the lab of Professor Roberto Dulbecco. Dr. Marguerite Vogt, who worked with him, had cultures of Chinese hamster cells and showed me how to culture and prepare HeLa cells for chromosome analysis.

"I used HeLa cells to confirm that tritium labeled DNA segregated in animal chromosomes, just as we had shown in plants. I published the results only as a part of a symposium and a Sigma Xi lecture series that I gave in the U.S. during the winter of 1960. HeLa cells, a subtetraploid of human cancer cells, have too many chromosomes for me to become very interested in using them on a regular basis....

"However, Vogt had two lines of Chinese hamster cells in the lab ... these hamster cells had maintained the diploid chromosome number.... I recognized that the Chinese hamster cells were ones that I would like to use for a variety of studies. I began by studying the sequence of replication within chromosomes and among different chromosomes of the complement, a project that I had begun earlier with root cells of the plant *Crepis*.... The most interesting differences involved the X and Y chromosomes.... The two X chromosomes in the female, one had very different patterns of replication.... The Y chromosome and the long arm of one X replicated late, appearing to carry inactive genes.... Later, I used the techniques learned at Cal Tech to collaborate with Grumbach and Morishima on studies of the time of replication of X chromosomes over the cell cycle in patients with aberrant numbers of X chromosomes ... late replication appeared to be correlated with gene inactivation in X chromosomes and perhaps in all chromosomes. The Chinese hamster cells remained a favorite in my lab for many years following my introduction to them at Cal Tech" ("My favorite cells with large chromosomes," *BioEssays,* 1991).

A series of Herb Taylor's experiments over the period 1958 to 1965 demonstrated the pattern of labeling of chromosomes over the cell cycle. It had been expected that, as in bacteria, DNA replication would start at one end of the chromosome and would proceed over time to the end. Taylor found something quite different. **Replication was occurring in many spots at once, and no preference for a particular end was evident for initiation or termination. In the same work he made an even more startling discovery—that one of the X chromosomes was always late replicating in mammals. Taylor is credited with making the first observation of the late-replicating X chromosome in mammals.**

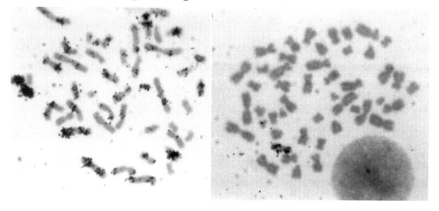

Autoradiogram of chromosomes from a human leukocyte which has been exposed to H³-thymidine for 10 minutes late in the S interval [(DNA replication)] and then transferred to a medium with an excess of unlabelled thymidine. The localized sites of synthesis characteristic of a number of chromosome [sites] are illustrated, in addition to the intense labeling over the whole length of the X- chromosome.

[When the cell was exposed to H³-thymidine at a later interval of S only the X chromosome is labeled (second photograph).] (Proc. Natl. Acad. Sci. USA, 1962)

He summarized this work in an address to a 1967 meeting, "Nucleic Acid Metabolism, Cell Differentiation, and Cancer Growth," including the new perspective of his 1966 chromosome model.

Taylor: "**The first evidence that the sequence of replication is highly regulated** was obtained in autoradiographic studies of the root cells of *Crepis* (Taylor, 1958) and somatic cells of the Chinese hamster (Taylor, 1960). The latter illustrates the point best. Established cell lines from male and female embryos were grown in vitro where H³-thymidine could be readily supplied to the cells for a short time and then removed and replaced by a medium containing unlabeled thymidine. Samples of the unsynchronized population were fixed at hourly intervals thereafter and the incorporation of H^3-thymidine into various chromosomes was examined in autoradiographs. It was clear that all parts of most chromosomes were not being labeled by such a pulse exposure to the isotope.

"The most striking variation in labeling large blocks of DNA occurred in the sex chromosomes, but was by no means restricted to these. For example, in male cells, the first labeled chromosomes to reach metaphase (those labeled at the end of the DNA synthetic phase), invariably had more tritium in the X and Y-chromosomes than in the autosomes. On the other hand, the last labeled cells to reach metaphase (those labeled in early S) had little if any label in the Y-chromosome or in the short arm of the X. In female cells labeled early in S, one X was almost wholly labeled while the other had tritium only in the short arm. If labeling occurred late in S in female cells, the whole of one X and the long arm of the other contained tritium. To summarize, in cells derived from a male the Y-chromosome and the long arm of the X were replicated in the last half of S, while the short arm of the X-chromosome was replicated in the first half of S. **In cells derived from a female the whole of one X and the long arm of the other was replicated in the last half of S, while only the short arm of one X was replicated in the first half of S.** Since then, this out of phase replication of the X and the Y-chromosomes as well as sectors of autosomes has been established in many different species of both animals and plants. That the pattern is consistent from cell generation to cell generation was indicated by autoradiography....

"The basis for this **correlation between late replication and genetic activity** is not yet understood, but the widespread occurrence of the phenomenon and its changes during differentiation suggest **that a major role in the regulation of gene action is involved**....

"Before we can make much progress in understanding the significance of these variations in replication and condensation, much more must be learned about the control of DNA replication at the molecular level and the actual mechanisms involved in replication *in vivo*.... On the basis of autoradiographic evidence, we have for a long time assumed that chromosomes consist of units of DNA that can be replicated independently. These hypothetical units have recently been referred to as replicons (Taylor, 1963,

1964). Only recently have we obtained evidence concerning their size in higher cells and suggestions concerning their structure. What is the size of the replicons and what features mark the beginning and possibly the end of a replicon. These considerations do not require that we know the arrangement of replicons within a chromosome, but much of the present evidence suggests that they are tandemly linked, perhaps with no non-nucleotide spacers as we had originally supposed. **A recent model based on the idea that a whole chromosome arm may be one continuous piece of DNA has been presented**" (Nucleic Acid Metabolism, 1967).

It was hard for anyone even to imagine what we now take for granted—that a chromosome could be one continuous unbroken piece of DNA. This was an astonishing idea when Herbert Taylor proposed it in 1966, and his experimental work was consistent with it. It was beginning to look as though the chromosomes of bacteriophages and animals were more alike than different.

In the same studies he describes above, Taylor made additional discoveries that weren't immediately appreciated by those outside the field. Taylor's studies of pulse-labelled chromosomes in synchronized cells revealed spatial relationships between individual replicating chromosomes.

Fellow cytogeneticist Sheldon Wolff: "… long before we had heard of chromosome painting with molecular biological probes, and of the term chromosome domains in interphase, Herb had shown by labeling the late replicating chromosome with tritiated thymidine that the label was not distributed throughout the interphase nucleus, as common wisdom held. We in cytogenetics fully appreciated this finding, which was later 'rediscovered' with the newer techniques" (letter, 1999).

Moving South—Canoes and Research

In 1963 Herbert Taylor published Part I of what became his three-volume series Molecular Genetics. *In the preface to that work he describes his reasons for putting this work together. It is also a good description of how his own scientific career was changing.*

Taylor: "Although genetics is one of the youngest subdivisions of bi-

Herb pulling canoe straight up out of Lake Nipigon, Northern Ontario, 1985.

ology, it has come to occupy a central position and … consideration … indicates that the emphasis in this work has shifted from the classical approaches of genetics to a molecular view of the mechanism of heredity; therefore *Molecular Genetics* is an appropriate title."

The work became a very popular graduate textbook, as well as a reference book for his colleagues in the field. The field was still so new that no other texts were available. Herbert Taylor had indeed become a molecular geneticist, but only a year later a new challenge tempted him away from the academic environs of Columbia and the harsh climate of New York.

Taylor: "By 1964, the inconvenience of my long daily commutes from the Westchester suburbs into New York caused me to leave my position at Columbia for the more livable surroundings of Tallahassee and Florida State University" (unpublished memoir, 1997).

Having produced a book on the origins of molecular genetics (Selected Papers on Molecular Genetics), Herbert Taylor had evolved into a molecular geneticist by the

Shirley and Herb Taylor in Tallahassee, 1974.

time he joined FSU's new Institute of Molecular Biophysics. Shirley was ready to return to an academic life, but the Florida nepotism employment rule prevented her from also having faculty employment, so she became staff manager of the lab, and students therefore enjoyed the benefits of working with two scientists in the lab. Several of his doctoral students at Columbia also came with Taylor to Florida to finish their degree work.

Opportunities for year-round outdoor activity were part of what finally drew the Taylors to Tallahassee. They carefully planned their dream house and had it built, nestled in a lush wooded spot near the university. It almost gave the feel of their earlier wilderness camping. The kinder climate and accessible Gulf coast provided a variety of ways to enjoy nature. Before long they learned that the best way to explore Florida swamps and wilderness coast was by canoe. For years they went canoeing most weekends, traveling nearly all of the fifty possible canoe trails within reasonable driving distance. At times they took along graduate students who were interested in exploring north Florida. Most forays gave Taylor some time to fish, an activity he had enjoyed when a boy near the Red River in Oklahoma. Herb and

Herb and Shirley Taylor with visiting grandson, canoeing and fishing in Florida's Aucilla River, 1991.

Herb fishing in North Florida, 1990.

Shirley continued to spend long days fishing from their canoe in the nearby St. Marks Wildlife Refuge's wilderness waters. It was there that they celebrated their 50th wedding anniversary, still pulling their canoe over the dike to fish in their favorite Gulf of Mexico saltwater bayou. It was not so much whether they caught keepers but that they had a day in which to savor the solitude and beauty of the watery wilderness.

Back to Meiosis

When Taylor began as an independent researcher, his plan was to study meiosis (the division that produces sperm and egg). The events at meiosis, along with the random pairing of gametes at fertilization, determine the genetic make-up of eventual progeny, so meiosis seemed more directly relevant to genetics than did mitosis. His goal had been to determine whether the chiasmata (crossed chromosomes) seen in meiotic chromosomes were really exchanges of information or simply artifacts with no genetic consequence. He had been temporarily distracted from this pursuit by his chance to demonstrate the semiconservative nature of replication in mitotic chromosomes, which had made an immediate contribution to current knowledge. The more difficult experiment in meiosis took several more years. His early years investigating meiosis and learning to culture anthers had prepared the road, and his dramatic work demonstrating incorporation of ^{3}H-thymidine in mitotic cells paved it for the work he would do in meiosis.

Taylor: "For the first four years after returning to academic life [after returning from the war], I spent all available research time, which was limited by a heavy undergraduate teaching load, on studies of meiosis in *Tradescantia* and other plants with large chromosomes. The diploid species of *Tradescantia* have 6 pairs, and another favorite, the Easter lily, has 12 pairs. I had decided to use the plant physiology I had learned with my thesis advisor for the master's degree, O. J. Eigsti, the codiscoverer of the effect of colchicine on cell division. Under his guidance I had produced polyploid phlox and made a study of some of the physiological changes. I hoped to culture anthers in vitro and thereby change the environment of the microsporocytes enough to produce desynapsis, or at least variations in chiasma frequency, that might reveal something about the regulation of crossing over in plants with large chromosomes. I failed in that objective, but was able to make good microscope preparations of prophase and later division stages from cultured anthers that were superior to any I could obtain from fresh material.

"I also began a concerted effort to study meiosis with the objective of understanding the mechanisms of crossing over, and to see if segments of homologous chromatids were exchanged during meiosis. My preliminary experiments at Caltech (sabbatical) and Columbia University had indicated that the problem would be difficult. I chose a grasshopper that could be purchased from commercial sources, the large 'lubber grasshopper', *Romalea microptera,* with 8 large and 3 small autosomes and an X chromosome of intermediate size that is heterochromatic in spermatocytes. Since inbred strains of grasshoppers had never been developed, we would have to use wild populations."

Florida's lubber grasshopper (photo by James Castner, University of Florida, IFAS).

In Florida the grasshoppers could be collected from the wild, freeing the lab from dependence on the commercial supplier. The entire family helped collect lubber grasshoppers each spring from the wild in the Everglades. Only thousands of autoradiographs over a period of 6 or 7 years revealed the precise injection times and timing sequences required to demonstrate sister chromatid exchanges in the grasshopper.

Taylor: "After moving to Florida, I collected each spring from the field, as the commercial suppliers had done. To find labeled meiotic cells, the nymphs had to be injected and sacrificed at intervals to see when labeled chromosomes appeared. The only synchrony that would help was that occurring in the cysts of the testicular tubules. Sixteen

Radiographs of dyads from cells which incorporated thymidine-H^3 one cell cycle before the premeiotic interphase (arrows indicate the terminal centromeres). [First panel] Dyad with no visible exchanges. [Second panel] Dyad with proximal reciprocal switch points for labeled and unlabeled segments, and a distal non-reciprocal switch point. [Third panel] Dyad with two non-reciprocal switch points for labeled and unlabeled segments. (J. of Cell Biol., 1965, Figs 10, 11, and 12).

spermatocytes are naturally synchronized in each cyst along the tubules that make up a testis. All stages from 2-celled cysts to spermatocytes could be found in a stage-dependent sequence along each tubule. In older grasshoppers, maturing sperms are found at ends of the tubules connected to the sperm duct. To get chromosomes entering meiotic prophase with one labeled chromatid and one unlabeled, a cyst of cells would have to be labeled at the interphase before the last mitosis that produces sixteen spermatocytes. These cells would reach metaphase I some weeks later. If the cells of a cyst were not in S phase during the time the injected thymidine was available, or if the available labeled thymidine spanned two S phases, the appropriate pattern would not be seen. We learned that an injection would supply thymidine for less than an S phase, but rarely all of the cells in a cyst would be appropriately labeled. The premeiotic S phase is longer than the previous S phase and therefore much more likely to be labeled.

"**It took a number of years and many difficult experiments to prove finally that crossing-over involved the physical exchange of segments in chromosomes during meiosis.** One has to be patient in such meiotic research because only one crop of nymphs was available each year. Once the appropriate labeled cells were found, it was then necessary to distinguish between genetic exchanges and sister chromatid exchanges that might have occurred during the two preceding cell cycles. I finally obtained appropriately labeled cells from a few cysts and determined that exchanges between homologous chromatids do, indeed, occur. **Nearly 10 years after the first demonstration of sister chromatid exchanges, *homologous* chromatid exchanges were finally demonstrated.** However, hardly anyone was interested compared to the excitement over the experiments I had designed in 1956 to demonstrate tritium labeling and the resolution that would allow a study of meiotic crossing over" (*BioEssays*, 1991).

Taylor: "After I finished the work on meiosis in grasshoppers, I spent one more summer [1966] on meiosis with H. G. Callan in St. Andrews where we worked on the spermatocytes of the common newt, *Triturus vulgaris*. Its large chromosomes and rather synchronized meiosis, we thought, might provide us with nearly ideal material for the studies of the correlations between chiasmata and physical exchanges of chromatid segments. I knew some of the problems connected with timing of labeling in relation to premeiotic S phase from my studies of the grasshopper, but I did not anticipate that the premeiotic S phase in the newt would extend over a period of 9–10 days. Callan later showed that the lengthy S phase is correlated with a small number of widely spaced origins of replication, unlike cells at other stages of the life cycle where S phase and origin distribution differs from that in mammals only as one might expect considering the differences in chromosome and loop size. We apparently did not inject early enough in the season to obtain chromosomes in spermatocytes with one labeled and one unlabeled chromatid. These large chromosomes with excellent meiotic stages were a good choice, but in one summer our objectives could not be reached.

"Many of the problems of meiosis will probably be solved with organisms with the smallest chromosomes, but my hopes and dreams of solving some of the problems of meiosis and the mechanism of crossing over that I began as graduate student have led me on an interesting odyssey" (*BioEssays*, 1991).

Taylor remained a respected figure in the community of meiosis research throughout his career and into retirement. His fundamental work of demonstrating the physical exchange of chromosome segments during recombination was of seminal importance in the field. In 1973 as chair of a session on meiosis at the International Genetics Symposium, he addressed the audience with a brief history of the theories of meiosis and genetic recombination. He ended with a warning that has been quoted in hundreds of papers since:

Taylor: "In spite of all these promising leads, three central problems which were delineated and appreciated many years ago are still with us, namely (1) the mechanism of homologous pairing in zygotene; (2) the mechanism of chiasma formation and crossing over; and (3) the basis of segregation, i.e., the affinity of sister chromatids after diplotene, the terminalization of chiasma and the affinity of homologous chromatids and the related manipulations of bivalents characteristic of first meiotic division.

"**I will only remind you that meiosis is still a potential battleground where dead hypotheses litter the field or rest uneasily in shallow graves, ready to emerge and haunt any conscientious scientist who tries to consolidate a victory for any particular thesis**" ("Meiosis," International Genetics Symposium, Introduction by Chairman; published in *Genetics* 78; 187-191, 1974).

Later, in a letter to a colleague and friend who applauded that presentation, Taylor wrote with his usual humility: "I have not really worked on meiosis for years.... But I always keep the problems of meiosis cooking even if on the back burner" (letter to Earlene Rupert, 1973).

Dreams of Travel Come True

While their children were small, the Taylors had postponed their wartime dreams of foreign travel. Herb made the lecture tours and trips to scientific meetings in the United States, Europe, Australia, Brazil, and Japan alone, but the couple did take their young family abroad for the summer of 1966. They drove through Germany, Switzerland, and France, camping in their Volkswagen bus and spending a week at the Max Planck Institute in Tübingen. Whether they slept near an ancient castle or heard their first cuckoo, every day of that month brought interesting experiences. The final destination was St. Andrews, a small town just north of Edinburg, where they spent the rest of the summer. The family remembers well the rented house and garden that the absent physicist owner had named High Entropy (for his random, not orderly, English garden!). The children enjoyed their first experience in that small town, safe for them to explore freely. They could walk alone to the Byre Theater (a converted cow shed seating 45!) for each new play and today fondly recall having seen Shaw's Arms and the Man *and* The Fantastiks. *Both Taylors were there to work on meiosis in the salamander with Professor H. G. Callan, outstanding cytologist and head of the Zoology Department at the University of St. Andrews. Although a paper resulted from the work, they weren't able to accomplish what they had hoped. They did, however, successfully program themselves that summer to arrive promptly for departmental tea at 11 and at 4.*

◇

Students—Mentoring All Along the Way

While living across 120th Street from the lab at Columbia, the Taylors frequently entertained the graduate students from the laboratory, introducing international students to American holiday dinners and reciprocally learning something of the traditions of other countries. The students frequently brought some of their native foods to share, much to the delight of their hosts. Recipes were shared and the Taylors added these exotic dishes of Indian, Chinese, and Egyptian origin to their own family repertoire, thus beginning a Taylor tradition. Both Shirley and Herb loved creating foods in the kitchen, preparing exotic dishes as well as interesting variations of ones from their own southern and midwestern American backgrounds. Generations of graduate students had their first exposure to international cuisine at a gathering in the Taylor home or an in-lab lunch.

Taylor: "My first doctoral student at Columbia University was a woman, Sandhya Mitra, who was later a postdoctoral fellow at the Rockefeller Institute in New York. She married an engineering student at Columbia University and they returned to India where he advanced to be the president of a leading university. Sandhya continued her research while rearing a family and is now a major interpreter of molecular and cellular biology in India. Of my 33 Ph.D. graduates, 13 are women. All but one or two have continued in research, teaching, or both. A number of other undergraduate and master's degree students have gone on in some area of cellular or molecular biology.

"Dr. Francis Ryan (in zoology) and I obtained one of the first training grants made by the NIH when I was on the faculty at Columbia University. In 1958 I had been made the first Professor of Cell Biology in the history of the university with a joint appointment in Zoology and Botany. We soon were training up to 12 to 15 doctoral students in cell biology and genetics. I was a hands-on cell biologist who worked in the laboratory with one research assistant. My students worked on related problems, but each was an independent investigator with a problem selected in seminars and discussions with me. I was called Dr. Taylor by my students even though our contacts were close and regular as their research advanced. When publications were made based on their research, my name appeared in an acknowledgment as an advisor rather than a co-author. However, through the years as our research became more focused and I had more direct input into the research of my students, we published jointly with me as the second author" (letter to Susan Gerbi, past president of ASCB, in 1993).

Taylor's first Columbia graduate, Sandhya Mitra, wrote:

Sandhya Mitra and Herb Taylor at his home, Tallahassee, Florida, 1994.

"Taylor represented several different classes of mentors and guides. At the academic level, his quiet insistence on nothing short of excellence, his examples to make us think objectively and not be diverted by tempting conclusions, and his non-imposing but persuasive ways of making us self-reliant molded, at least myself, into a somewhat similar teacher" (letter, New Delhi, 1999).

And in letters sent to Shirley after Herb's death, many of his graduate students described him as teacher and mentor. Almost 40 years of training young scientists—and through all those years his characteristic love of, and excitement about, discovery, his pleasure in sharing it with new scientists, was obvious to each of them.

Barbara Boyes: Certain things I admired in Dr. Taylor I have tried hard to develop in myself; particularly the clarity, conciseness, and logical organization of his writing, the sheer fun of working in the lab, and the careful way in which an idea is worked out and vigorously tested. I am very grateful for the excellent start he gave me in the exciting and demanding career of scientific research."

Bill Haut: "I remember his eyes bright and sparkling with anticipation as he searched the data for an explanation. And I remember him, glasses in hand, eyes tightly closed, quietly searching his imagination for questions as much as answers. Like an artist with an empty canvas, he kept painting new pictures with searching creativity, talent, and a willingness to risk. How fortunate I was to know this man both as mentor and as friend."

James Pollack: "Doing research in Herb's lab was an adventure not only in research science, but also in scientific attitude and the curiosity that drives it. Herb was always open to new ideas and avenues of research. It made no difference if the idea came from a graduate student or a peer professor; Herb always gave the idea his genuine consideration."

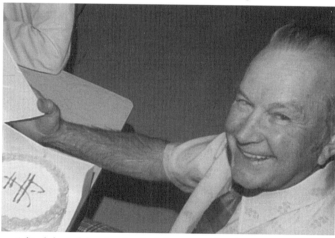

A birthday celebration in the laboratory, Florida State University, Tallahassee, Florida.

Tommy Laughlin: "But it was really the example he set that led the way in my life that followed graduate school. That example went far beyond the lab work and dissertation writing. He showed me a positive exciting approach to life. He provided me with a vision of how things can be done when you combine knowledge, drive, patience, and compassion."

Barbara Boggs: "The true magic of being in his lab was not just that he had a high standard of excellence but that he also cared about people, for who and what they were outside the lab. To have stumbled into Dr. Taylor's lab as an undergraduate was one of the luckiest things that has ever happened to me. Dr Taylor will always be my role model in science and in life."

Lynne Wall: "Herb struck that balance between being highly supportive on all fronts but allowing at the same time independence and a sense of responsibility to predominate in the lab and in all matters to do with research."

Marcia Applegate Harris: "I remember he was the perfect teacher for me. He fed me ideas and let me grope my own way to a project, never critical of my constant meandering off the main track."

Joan Hare: "These qualities I have inherited from Herb Taylor, and for these I am extremely grateful — the love of discovery, devotion to the search, excitement in the design, the ability to make it yourself and fix it yourself, and patience. These have served me well in the life that followed graduate school.

He taught us not only science, but real life as well.... He taught us to deal gently with people."

Taylor at home on Hilltop Drive, Tallahassee, Florida, 1985.

Chapter 4

Models of Chromosomes

As a student, Herb Taylor had studied classical ideas about the nature of chromosomes. He first published as a graduate student working with O. J. Eigsti, on the induction of polyploidy (doubling or quadrupling of the chromosome number) using colchicine. This work became a classic in genetics and was used later in commercial plant breeding as the basis for production of many polyploid seed stocks. During his student years he also studied and published work on the physiological differences between diploid plants and their tetraploid relatives. When he began teaching genetics himself, he found it challenging to explain genes without knowing the physical nature of chromosomes. He focused his research on defining the chemical nature and behavior of chromosomes.

Taylor's groundbreaking demonstration of the semiconservative replication of DNA disproved most previous models of chromosome structure and forced scientists to consider new ones. Over the years Taylor became a major force in the development of new models that attempted to explain and reconcile all available experimental data. These models were instrumental in directing the course of research in the field. At the same time, Taylor was a major contributor to research on DNA replication and chromosome behavior. His models included data from many different laboratories, including his own. This work revealed yet another of Taylor's great talents, his ability to integrate data from disparate sources and synthesize cogent and coherent models.

Taylor's chromosome models recount a history of the molecular biology of chromosomes and DNA, from 1957 to 1969, a period of rapid growth of knowledge and change in ideas. Although the early models appear strange to our eyes today, by 1969 the model had evolved into something that should be recognizable to genetics students of 2004.

For years no one envisioned that a chromosome could be composed of a single long piece of DNA. The early models included some type of linker between shorter pieces of DNA. One of Taylor's early models proposed a tRNA-like linker; years later some viral chromosomes were discovered to have t-RNAs covalently attached near the end of their DNA. One of Taylor's later models had protein linkers, and still later models used proteins as a scaffold that held the long loops of DNA in place. Today we know that scaffolding proteins do just that.

One of Taylor's former students from the 1960's (Terry Ashley) remembered the affectionate joke his students had about their mentor: "Taylor's models are like automobiles—there's a new model every year!"

Multistranded Model

The first Taylor model to reach a wide audience was one he introduced in his 1957 paper in the Proc. Nat'l Acad. Sci. USA *along with the autoradiograms that demonstrated semiconservative replication in mammalian chromosomes.*

Taylor: "… the chromosome is several orders of magnitude larger than the proposed double helix of DNA. We know that these large metaphase chromosomes are coiled into at least one helix at the microscopic level and perhaps are twice coiled, a helix within a helix. That the chromosome could be a single supercoiled double helix of DNA is inconceivable when one considers the amount of DNA in a large chromosome. Chromosomes are much more likely to be composed of multistranded units. To explain their duplication as well as their mechanical properties at the microscopic level, they may be visualized as two complementary multistranded ribbons lying flat upon each other. Ribbons of this type with more flexible materials on their edges have a tendency to coil. If the edges contract faster than the central strands when the chromosome begins to shorten, the ribbons fold, one within the other, so as to form a long, trough-shaped, cylinder and with further contraction they assume the shape of a helix. Continued differential contraction would produce a helix within a helix...."

Taylor's idea of differential contraction resembles the supercoiling of DNA with which we are familiar today. It not only could explain the apparent coils visible in some metaphase chromosome preparations, but could also help to explain the enormous packing problem that chromosome biologists would continue to wrestle with for years.

"Although the chromosome model is provisional and may require considerable modification and refinement, it has many features necessary for duplication and the known stability of genetic materials. The large surface area exposed when the two complementary ribbons are extended would facilitate their rapid duplication. A double-stranded unit with two complementary faces has a high stability and if the two complementary units are composed of multiple identical strands cross-bonded, the stability of large units should be even greater. Such large units would

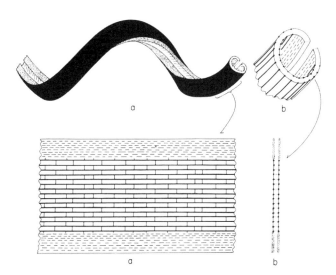

MULTI STRANDED MODEL *Upper panel: Schematic drawing of the proposed ribbon-shaped chromosome with the two multistranded units folded together and coiled; a, a single gyre from the coiled chromosome; b, detail in cross-section.*
Lower panel: Diagrammatic sketch of the multistranded units uncoiled and flattened; b, cross-section. The number and size of strands shown have no special significance. Although the assumption is made that the strands contain DNA, they do not necessarily correspond to Watson-Crick double helices (Proc. Natl Acad. Sci. USA, *1957, Fig 4).*

have a high probability of being transmitted as physical entities. If separation of the complementary faces involves the separation of intertwined double helixes of DNA, unwinding presents a problem, but perhaps not an impossible one" (*Proc. Natl Acad. Sci.* article, Fig. 4, 1957).

In a review years later, Taylor wrote:
"It seemed to many of us in 1956 that a chromosome must consist of many strands of DNA and protein interwoven into a complex structural fabric… " (*Genetics and Developmental Biology*, 1969).

Centipede Model

By the following year, Taylor's model of a chromosome had changed considerably. He added the idea of a protein backbone for attachment and stability. This model soon picked up the nickname "centipede" model, because it resembled a drawing of that creature. In his book Meselson, Stahl and the Replication of DNA, *Frederic Holmes describes Max Delbruck as being captivated by Taylor's work when he came to Caltech to discuss it in January of 1957.*

Delbruck: "Your model of chromosome structure, with a double backbone and DNA chains attached to both sides, continues to be discussed widely. It has very attractive possibilities beyond those we discussed when you were here. I would urge you to propose it soon (We call it the centipede model.)" (letter to Taylor, February 1957).

Taylor explained the modified model first for a general audience, in Scientific American, *following Watson and Crick's initial publication in* Nature. *He described clearly all the logical reasons it seemed impossible that a chromosome could consist of only a single strand of duplex DNA.*

Taylor: **"Our picture of the chromatid as a two-part structure fits very well with what we know about the DNA molecule and with the Crick-Watson theory. DNA too is a double structure, consisting of two complementary helical chains wound around each other.**

And some of our recent experiments indicate that the two strands of a chromatid are complementary structures. It is tempting, therefore, to suppose that a chromatid is simply a chain of DNA. But when we consider the question of scale, we realize the matter cannot be so simple. If all the DNA in a chromatid formed a single linear chain it would be more than a yard long, and its two strands would be twisted around each other more than 300 million times! It seems unthinkable that so long a chain could untwist itself completely, as the chromatid must each time it generates a new chromosome. Furthermore it has the wrong dimensions to be a single DNA chain. When fully extended, it is about 100 times thicker and only about 10,000th as long as the linear DNA chain would be.

"Under a high-power microscope we can see that the chromatid is a strand of material tightly wound in a helical coil. But beyond this the optical microscope cannot resolve details of the chromatid's structure.

"We know that chromosomes contain protein. So as a start we may picture the chromatid as a long protein backbone with DNA molecules branching out to the sides like ribs. Because the chromatids split in two, we visualize the backbone as a two-layered affair whose layers can sepa-

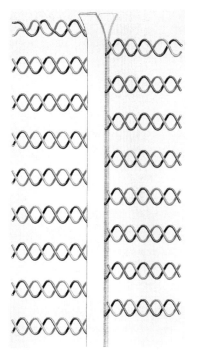

*[**SIDECHAIN** or] **RIBBON MODEL** of a chromatid consists of a two-layered central column to which DNA molecules are attached. One chain of each molecule is anchored to the front layer and the other to the back layer. When the chromatid duplicates, the central ribbon peels apart, unwinding the DNA molecules. Each half of the structure then builds itself a new partner* (Scientific American, *1958*).

rate. The ends of the two strands of a DNA molecule are attached to these layers: one strand to one layer, the complementary strand to the other. As the layers peel apart, they unwind the strands. The unwinding strands promptly begin to build matching new strands for themselves. Eventually the new strands also assemble a new backbone, and the original chromatid is thus fully duplicated" (*Scientific American*, 1958).

Ribbon Model

In subsequent papers, Taylor developed the model further, expanding it to describe the mechanisms of separation of the "ribbons" at division.

Taylor: "A model suggested earlier will serve to explain our results and in addition possess many intriguing genetic implications. We shall retain the essential ribbon shape suggested, but shall assume that the DNA double helixes are attached to a ribbon-shaped central axis or core which is a duplex (a double ribbon). The DNA double helixes are attached in such a way that one polynucleotide chain is attached to one ribbon of the duplex and the other polynucleotide chain to the other ribbon. Duplication can be visualized as beginning at the end by separation of the two polynucleotide chains in the region adjacent to their attachment points. This would initiate the unwinding of the double helixes, and replication of the DNA could proceed as proposed by Watson and Crick. With one end of each double helix free to rotate and with single bonds that allow rotation along each polynucleotide chain in regions where separation is occurring, the unwinding of relatively short side chains of DNA should present no mechanical problem. As the two ribbons of the core separate, we may visualize the assembling of two new ribbons, one along each of the original ones to form two new duplexes. All the new polynucleotide chains would now be attached to the new ribbons and would therefore act as two units in future duplication. Only by exchange of segments along the axis (sister chromatid exchange) would the new units be broken up. **The model is merely a skeleton of a chromosome, and only enough detail is introduced to explain our data. A number of questions immediately arise concerning the position of other components believed to be**

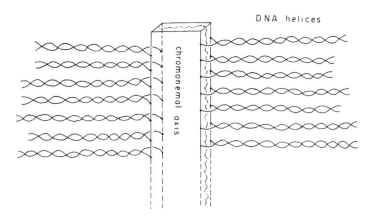

RIBBON MODEL *Model of a chromosome proposed on the basis of the behavior in distribution of labeled DNA. Side branches represent DNA double helixes (*A Symposium on Molecular Biology, *1959, Fig. 5).*

present in chromosomes and the nature of the hypothetical core. Speculation on these matters is probably not fruitful at this time. The critical test of this model will probably come from data on genetic recombination, which have always been the most powerful tool for the analysis of chromosome structure (*A Symposium on Molecular Biology,* 1959).

Ladder Model

The objection to the ribbon and centipede models was that the DNA duplexes had free ends, which although free to rotate, were also not in a fixed linear order. That was difficult to reconcile with decades of genetic data. Somehow the chromosome needed to be reconciled with a linear map. The first attempt produced the ladder model, in which the free ends were joined. Taylor credited Ernst Freese with first suggesting this idea to him. The first version of this model was introduced in the 1958 Scientific American article as an alternate model. A more developed version appeared in the 1962 International Review of Cytology *article and later in his 1963 textbook,* Molecular Genetics, Part I.

Taylor: "**Models are useful only in so far as they allow us to explain data, to bring order into our thinking, and to predict and plan significant experiments. Such models should be as simple as is consistent with these requirements at any given stage in our process.... If one simple model will explain all of the data, its detailed description is probably worthwhile although such a model will seem inadequate to some investigators and, since we certainly do not have all the information, it can at best be only partially correct.**

"Any satisfactory model must have the following features: it should 1) contain two sub-units of DNA organized in a way that will explain the pattern and frequency of twin exchanges; 2) allow separation and sorting of the two chains of the DNA double helix during or after each duplication; 3) explain how a chromatid can appear morphologically double, yet act as a single unit in recombination, a double structure to radiation during division stages, and then become single in reaction to the same agents after division; 4) explain the various morphological appearances of multistrandedness and the contraction and coiling observed with the light microscope; and 5) explain how a change in a single base pair of DNA might be expressed immediately as a mutation.

"Traditional cytological models are of very limited usefulness in explaining most of these features. On the other hand, the acceptance of a chromosome model consisting of a single long double helix of DNA is conceptually difficult. In an attempt to find a solution a side chain model was suggested. It consisted of a double protein core to which relatively short segments of DNA were attached.... The first deficiency in this model was revealed when it proved inadequate for predicting the pattern and frequency of sister chromatid exchanges [but with some modifications this could be explained].

"In addition, genetic recombination, although inadequate to rule out a side chain model, indicates a linear order down to the smallest measurable unit. A variant of the original side chain model [ladder model] places the two halves of the protein core on opposite sides with nucleohistone molecules lying between like rungs on a ladder. As in [the ribbon] model a polynucleotide chain would be attached to only one axis through a bond that would allow rotation. Its complementary chain would be attached to the opposite axis. Freese suggested the important variation that the axes should be interrupted between alternate DNA helixes on each side of the ladder. When the alternations are properly alternated on the two sides, the result is a linear array of DNA segments linked tandemly. These linkers will be called 'R' linkers because they play an important role in recombination.

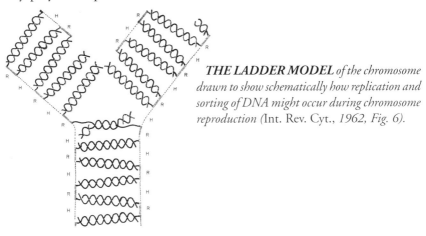

THE LADDER MODEL *of the chromosome drawn to show schematically how replication and sorting of DNA might occur during chromosome reproduction* (Int. Rev. Cyt., *1962, Fig. 6).*

"Having supposed interruptions in the axes it was then necessary to assume the sites to be rather stably bonded, at least during replication. These transient linkers will be referred to as 'H' linkers since they will be assumed to be important in the formation of half-chromatids. The chromatids then may conform to not one but to several models at different stages in the cell cycle. During DNA replication, and the condensed stages of division, it will be assumed to be in the ladder form. Following DNA replication and separation but before new 'R' linkers are formed, each chromatid will be in the side chain form with a single axis composed of alternating 'H' and 'R' linkers (*International Review of Cytology,* 1962).

The Power of Models

Taylor had already developed several models when he wrote and published Molecular Genetics *in 1963, the first text in the field, which served as both a reference for investigators and a graduate textbook. In the chapter he authored, he talked about the power of models to influence a field.*

Taylor: "Models which allow us to explain data, to express more concretely certain concepts, and to plan and design future experiments are useful. However, models have the disadvantage that they tend to freeze our thinking in such a way that significant data may be neglected. When we are fully aware of such disadvantages models can be useful, even when the information for making them is limited. Several recent attempts have been made to construct models of a chromosome (Taylor and others, 1955-1961). All of these are useful in crystallizing concepts, but all are deficient in other respects.

"Any model of a chromosome based on the present information should be composed of one, or at most a few, DNA double helices that extend the length of a chromatid. These must function as two subunits during replication at all levels of organization.… However, in chemically and radiation-induced breakage and exchanges, as well as reciprocal recombination and sister chromatid exchanges it acts as a single unit except during prophase. On the other hand, the pattern and frequency of sister chromatid exchanges indicate that even though the exchanges are only between whole chromatids, each chromatid is composed of two unlike subunits. When all of the evidence is considered, the basis for 'twoness' would appear to be the complementary chains of DNA double helix with some tertiary structure added" (*Molecular Genetics, Part I,* 1963).

The evidence was accumulating and Taylor, along with a few others in the field, was beginning to reconcile himself to the idea that chromosomes are a single DNA duplex, despite the enormous packing problems this configuration proposes. In the same 1963 text, he lays out the problem.

A More Sophisticated Ladder Model

Taylor: "A model based on a single DNA double helix which extends through the length of a large chromosome without interruption has some of the features required by our present data, but other features would appear to be missing. One of these is the packing or folding of several centimeters of DNA for manipulation within the dimensions of a cell. The

DNA or 'chromosome' of phage T4 is more than 50 times as long. The chromosome of *E. coli* would be perhaps 50 times this length. Some of the largest chromosomes might be more than a meter in length. The most useful models to visualize the folding and unwinding of such a long piece of DNA are those based on a suggestion by Freese. The essential feature of the [ladder] model is a DNA double helix with a regular sequence of linkers alternating in the two chains and located opposite a gap in the complementary chain.... Linkers have not yet been identified in analyses of DNA and indeed this can hardly be expected.... A little speculation on the nature of the linkers may be useful in the interim.... Since protein synthesis appears to play a role in the reunion of chromosomes, the H linkers for the present will be assumed to consist of proteins.

"In addition to a capacity for folding and efficient packaging ..., a chromatid must have many growing points for replication, yet be closed for replication for long periods of function. The initiation of and control of replication would appear to require operator sites to explain the observed control in order or sequence and to prevent more than one replication at each locus (*Molecular Genetics, Part I*, 1963).

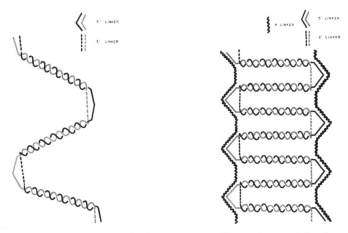

LADDER MODEL *A portion of a chromosome preceding replication. The plectonemically coiled strands represent molecules of DNA joined in tandem array by means of special sites or linkers in each polynucleotide chain. These linkers are assumed to occur in pairs at sites in which there is a reversal in polarity so that two 3' and two 5' ends of chains are linked by 3' linkers and 5' linkers, respectively.*

[Second panel] Stabilization by folding of the chromosome and the establishment of H-linkers is visualized as the step which precedes reproduction of each region" (Molecular Genetics, 1963, Part I).

Taylor went on to propose the existence of two types of linkers, one of which—the R linker—was a potential replication-initiation and recombination site and the second—an H link—the packing linker that completed the sides of the ladder, making the rungs parallel. 3' and 5' linkers actually joined the DNA strands, but the R and H linkers formed the sides of the ladder.

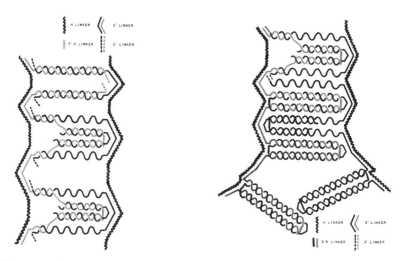

"During replication the 3' linkers are assumed to be opened by appropriate enzymes. The two subunits of the chromosome consisting of the alternate 5' linkers and the H linkers with the attached DNA chains are then held together only by the hydrogen bonds in the DNA. Two chains of adjacent molecules are assumed to begin replication by the insertion of a new 5' linker (Molecular Genetics, Part I, 1963)."

Taylor: "Very little can be said about the nature of the hypothetical 3', 3'R, and 5' linkers. The H linkers have been assumed tentatively to be peptides, but it now appears the other linkers cannot be peptides. [Some evidence] suggests that the 3' linkers or the 3'R linkers might consist of two phosphoserine residues coupled to the terminal nucleoside of DNA chains by an ester linkage in the same way that amino acids are coupled to transfer RNA. A diphosphate bridge could then couple two chains with a reversal in polarity" (*Molecular Genetics, Part I, 1963*).

Taylor: "**Of course, there are many possible non-nucleotide linkers that might provide the multiple sites for replication to begin.... The hypothetical linkers are designated by their assumed function**

rather than by their chemical nature. The chromosome is visualized as a tandem-linked series of 'replicons', segments of DNA which replicate as units (Jacob and Monod, '63), coupled by 3'R and 5'R linkers (replicon linkers) in such a way that there is a reversal in polarity in each chain at the junctions. During the condensed stages of division in large chromosomes the long strand is assumed to be coiled, folded and stabilized by H linkers that may give the chromatids a temporary doubleness during division stages.

"The model will serve as a basis for discussion of results. Since at present there is no way to test the model directly, its usefulness in predicting and interpreting results may serve to indicate its reality or to suggest necessary modifications" (*J. Cell. Comp. Physiol.*, 1963).

In this model, Taylor began to assign additional roles to the linkers; perhaps they regulated the initiation of replication and of recombination, in addition to providing a structure to the chromosome and thereby assisting in the equitable division of DNA between daughter cells.

◇

Replicon Model with Replication Guides

New data were published between 1963 and 1966 that did require the modification of the model as Taylor had predicted. Enzymes were discovered that made incisions in the DNA. The model no longer needed special-function linkers. Taylor unveiled his new model first in Part II *of the* Molecular Genetics *text, 1966.*

Taylor: "A chromosome model which is a modification of earlier models [is] described. In this model the DNA of a chromosome arm is assumed to consist of one long uninterrupted DNA double helix. No linkers are required since it is assumed that the duplex is opened by enzymatic scission at rather specific sites during replication. At some time following replication the sites are assumed to be closed by an enzymatic process similar to those which function in producing the covalent linkages of DNA chains during repair and recombination in phages. Non-nucleotide linkers between replicons would not change

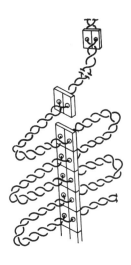

REPLICON MODEL *Diagrammatic model of a segment of a chromosome. The model shows only the DNA, a single duplex for each chromatid, and the hypothetical "replication guides" which are assumed to be coupled into a flexible column or axial element. Other components are not shown, but it may be noted that a prominent feature is a series of loops, which represent the units of replication. Even though a whole arm might consist of a single DNA duplex, one chain is assumed to be broken during replication on each side of a replication guide and then rejoined following replication* (Genetics & Developmental Biology, *1969,* Fig. *1).*

the model significantly, but since the evidence for any type of linker is meager, such a structure is not included in the model. The only hypothetical structures are proteinaceous components attached in pairs along the DNA duplex. One unit of the pair, a replication guide, is assumed to be attached to each chain of the DNA. These are assumed to be capable of coupling, which orients the chromosomal DNA into a structure of reasonable length and throws it into loops alternating on the two sides of an axis. The loops are assumed to be relatively longer than shown in the diagram. When the loops are extended a 'lampbrush' chromosome is formed, but when a chromatid is condensed as in division stages, the DNA of a loop is supercoiled to form a smaller ring. All such rings orient in planes nearly at right angles to the long axis of the chromosome. The two ranks of rings give a chromosome or a chromatid a nearly cylindrical shape especially when the axis is twisted and the two ranks of rings appear relationally coiled. The apparent doubleness of anaphase chromosomes could be explained by the two-ranked profiles of rings. The doubleness would be exaggerated by any agent or treatment which swells or unfolds the axial element presumed to be formed by the replication guides.

"Replication is assumed to occur independently in each replicon which is bounded by two pairs of replication guides. An assumption made in describing the model was that the DNA in the region of the replication guides might be a repeating sequence polymer of which there would be a few species characteristic of any particular type of cell. This DNA might be produced either at the replicator locus or outside the chromosome and would consist of single chain pieces of primer DNA with attached replication guides. These combined units which will be designated initiators, a term adapted from Lark, could recognize characteristic sites and could function in regulation of replication of the various replicons" (*Molecular Genetics, Part II,* 1966).

Taylor goes on to describe a scheme using the proposed primers for the initiation of replication from initiation sites. The replicon model presented in 1966 is remarkably close to our present day understanding of a chromosome. What Taylor called a replication guide in 1966 is now known to be a complex of DNA replication related proteins, and the entire structure is generally referred to as a replisome. The model correctly predicted the occurrence of sites where single-stranded RNA primers are made and DNA replication begins, and Taylor himself showed that DNA replication is initiated at many sites throughout the chromosome, so models no longer need specialized linkers to perform these functions. And what Taylor called 'replication guides' explains fairly well the function of replication proteins, which recognize both specific DNA sites on the chromosome and short RNA primers, regulating where and when replication intiates in a chromosome. Using supercoiling to explain the packing phenomenon solved another part of the difficulties of earlier models.

Taylor: "Note should be taken that it still seems necessary to assume that something holds the long chromosomal DNA strand in shape. These structures may be proteins attached to the DNA but do not necessarily interrupt its continuity. In the model and in a hypothetical scheme of reproduction which has been proposed (Taylor, 1966) these units are called replication guides. They are assumed to be attached to some sort of flexible column which is probably attached to the centromere. If the loops were supercoiled and disposed around a column in the metaphase chromatid, it would consist of a cylindrical rod similar to that seen in the light microscope…" (*Genetics and Developmental Biology,* 1969).

This was the final Taylor model that was intended to describe the struc-

ture of the chromosome. Later models explored DNA replication—initiation and extension and how different sites of initiation might be constrained to come into use at different times. Still later Taylor models attempted to explain the role of DNA methylation and the consequences of DNA mismatch repair.

Chapter 5

DNA Replication—from the Chromosome to the DNA Molecule

The use of tritium-labeled thymidine that Taylor had pioneered turned out to be equally useful in directly tracing the sequence of events during the replication of DNA. Instead of inferring what had happened in the DNA from the appearance and position of label in chromosomes, Taylor and his group began to study DNA isolated from the nuclei of cultured cells following a pulse of label to the cells. The cultured cells were much easier to manipulate, and the short pulses of label possible for cultured cells allowed a more detailed examination of the sequence of events in replication. He had learned the methods of density centrifugation as an analytical tool during his sabbatical at Cal Tech with Max Delbruck and realized this was a tool that would allow him to collect more information than autoradiography alone. The transition from labeled chromosome to isolated labeled DNA was a natural extension of his earlier work. In a 1969 paper he described how his results from autoradiography of labeled chromosomes led him to new questions that could be answered by the technique of density centrifugation.

Taylor: "Chromosomes consist of two DNA subunits before replication. One subunit is transmitted intact to each daughter chromosome with the exception that sister chromatid exchanges occur with a relatively low frequency, about one per division cycle for larger chromosomes. However, measurements of rates of chain growth in an established cell line from Chinese hamster indicated that there are several hundred replicating units per chromosome (Taylor 1968). The question raised is as follows: how are so many replicating units arranged in a chromosome, and how are their replication and integration into the chromosome regulated? Let us first consider briefly the evidence for multiple growing points. For example the large arm of the X-chromosome which replicates late in the S-phase spends less than three hours in replication. Yet it contains enough DNA to form a Watson-Crick type double-helix about 50,000 microns long. The new chains grow at about one micron per minute (Taylor 1968 and evidence presented here), and therefore pieces no longer than 180 microns could grow in the time allowed for replication. This means there would have to

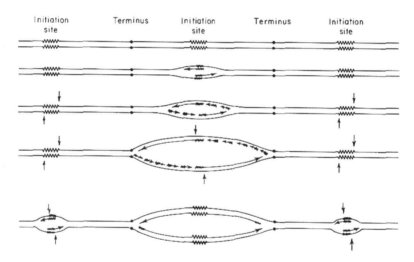

A model showing initiation sites for replication in chromosomal DNA of vertebrates. The spacing of initiation sites and termini has been indicated by limited melting (Evenson et al., 1972) by the size of **nascent fragments isolated from pulse-labeled DNA** (Taylor et al., 1973a) and by the **size of enzymatically induced fragments** isolated from stationary phase cells (Taylor et al., 1970). **Only the 2 μm segments** are assumed to be composed of 5 segments composed of 900 nucleotides of unique DNA with 300 nucleotides of intervening repetitive DNA (Davidson et al., 1972). Nicks are introduced and closed to allow unwinding and to explain the isolation of **Y-shaped replicative segments** by limited melting without shear (Taylor, 1973). The model is only detailed enough to account for known features of replication with the exception that continuous chain growth shown for one complementary chain in each unit has not been demonstrated; both may be replicated as short segments and joined by ligase activities (Inst. Rev of Cyt. 1974).

be 275–280 growing points working simultaneously and continuously to double the DNA in the given time.... " (Taylor, presented at the XII International Congress of Genetics, 1969)

Enthusiasm was high for this new area of research. Charles Thomas, a colleague at Harvard wrote Taylor, " Your whole visit was a pleasure and I felt I learned a great deal from our mini-seminars. If our respective results hold up, the problem as I see it, is to understand the relationship between the subunits and the slaves. I am tremendously enthusiastic about learning as much about higher chromosomes as we know about viral chromosomes. We are hoping to use your method to isolate subunits from trout sperm DNA." (C. Thomas, letter to Taylor 1969)

Taylor in his office, Florida State University, Tallahassee, Florida, 1980s

Taylor had found some of the first evidence that DNA replication in animal cells begins at many sites and that newly replicated DNA appears first as very short pieces not attached to the chromosomal DNA. When the cultured cells were treated so that all the cells in the culture were beginning DNA replication at one time, he could separate these small fragments from the rest of the chromosomal DNA. He could show that virtually all the tritium label from thymidine delivered to the cells in a brief pulse just before isolation was in these fragments. And on the basis of its susceptibility to degradation and different heating conditions, he could determine that all of the label was in RNA, which, with increasing periods of synthesis following the pulse of label, were joined to DNA. These short fragments were soon recognized as the primers of DNA synthesis and a critical first step in the initiation of DNA replication. The results were reported in a series of papers he published almost simultaneously with the work of other groups reporting similar results in animal and bacterial cells (Taylor papers, 1973-1977).

Geneticists working with DNA replication in bacteria frequently found answers before their colleagues working with eukaryotes,[10] but the complexity that awaited the explorers of the eukaryotic world was amazing time after time. Although the origins of DNA replication turned out to behave much the same in prokaryotic and eukaryotic cells, their organizations and thus their controls turned out to be much different. Taylor described both his own findings about the organization of DNA origins and also an idea about how they might be regulated in a review article within his third volume of Molecular Genetics, published in 1979.

Taylor: "It is well known that chromosomes of eukaryotes replicate one time each cell cycle. Mistakes are rare or non-existent. Yet chromosomes initiate from many sites, at which growth proceeds in both directions. Reinitiation does not occur until all segments of all chromosomes are complete. This process must be carefully regulated despite the presence of thousands of initiation sites in each cell.

"Recent evidence from studies of Chinese hamster ovary cells which were highly synchronized indicate that there are many more potential origins in cells than we had previously suspected for cells from the adult animal (Taylor and Hozier, 1976; Taylor, 1977). By holding cells synchronized by mitotic selection at the G_1-S phase interface by completely depriving them of thymidine, it was possible to show that many more of the potential origins in a limited part of the DNA could be activated, or made ready for use when replication was initiated by adding [^{3}H] thymidine to the medium. Fiber autoradiographs, which show the position of individual labeled loops which have been stretched into single fibrils by flow, indicate that many more origins

---

[10] *eukaryotes: organisms whose chromosomes are contained within a defined nucleus.*

are used.... The origins, estimated by measuring center to center distances between labeled segments, were from 4 to over 100 μm apart. When these distances were plotted in a frequency diagram the most common distances between origins were found at intervals of 4, 8, 12, 16, and perhaps 20 μm. This observation led to the hypothesis that potential sites in these fibroblasts, and probably all cells of the hamster, exist every 4 μm, the common integral for each interval measured. Not all the potential sites could be made active by the treatment, but the number actually used after holding cells for 20 hours after division, i.e., blocked at the G_1-S phase boundary for 12–14 hours, was about one in three, compared to one in 15 to 20 for cells not interrupted in the cycle.

"Not only the used origins, but the unused origins must be modified so that initiation is excluded during the S-phase involved.... Differences in methylation [would] provide a fail-safe mechanism that is extremely simple. Let us suppose that the replication complex binds at a fairly large palindromic sequence [only in the fully methylated state] which includes one or more small palindromic sites on each side of the larger palindrome.

"If methylation of these sites is delayed after replication until G_2, the site is modified automatically by the [replication] fork moving across the site. No other assumptions are necessary other than to suppose the replication complex cannot bind to a half methylated site" (*Molecular Genetics, Part III,* 1979, Chapter III).

This molecular model of the control of initiation of DNA replication in the chromosome exemplifies Taylor's passion for discovery. It was a completely new and innovative hypothesis to explain how the thousands of individual replication origins in eukaryotes could be marked so that they were used only once during a round of replication. Again we see that Taylor was a scientist who loved to explore ideas—how might this work? Here is a wonderful example of his scientific approach, formulating a question, proposing a theory, planning an experiment, analyzing the results, adjusting the theory to meet new data, and so on. But in one way he was different from many other scientists and perhaps more typical of his time; he was willing to put these ideas into print and share them with others before they had been tested. He was willing to let other people test his ideas. He was willing to be found wrong, in order to move closer to finding the truth.

◇
Why Is There a Methyl Group on Cytosine?

It was a puzzling phenomenon: the methyl group was added to DNA after replication in mammalian cells, and it was added only to cytosines that were followed by guanosines. What did it mean? Herb Taylor was one of a small group of molecular geneticists at the time who saw the phenomenon of methylation of DNA as a possible mechanism for control of genetic events. The particular genetic events seemed limitless—DNA replication, embryonic differentiation, DNA repair, tissue growth, development.

Taylor: "In 1977 I wrote a grant proposal in which I applied to study developmental patterns in enzymatic methylation of DNA in eukaryotes. One part of the proposal was to assay cells at different embryonic stages for maintenance and de novo type methylase activity. With one exception, the referees, probably all developmental biologists, recommended that the work not be supported because there was no evidence that methylation plays any role in eukaryotic gene regulation. Aside from proving that innovative ideas can seldom be used to successfully compete for grant funds, the skepticism of biologists toward methylation as a regulatory mechanism was, and still is, widespread even among those who investigate the problem. That is a healthy situation for all points of view should be brought to bear on a problem of such importance. However, to deny funds to investigate a problem because one has already formed an opinion is hardly commendable. The great skepticism about the significance of methylation is based in part on the evidence that it is absent or little used in *Drosophila*, a favorite organism for genetic and developmental studies. There now remains little doubt that methylation of cytosine in certain CpG (the two base DNA sequence of cytidine followed by guanosine) sites can strikingly effect the transcription of sequence 3' to the methylated doublet. How this inhibition operates and to what extent it is used in cells is still debatable. Furthermore, a mechanism for the inheritance of a pattern of methylation once it is installed is understood and demonstrated in principle.…

"I have concluded that DNA methylation is an important mechanism of cellular differentiation in vertebrates, although it is probably not in insects such as *Drosophila*. Even in vertebrates, methylation is only a subsidiary mechanism. The primary mechanism is based on a scheme that eukaryotic cells invented hundreds of millions of years ago and cell biologists and geneticists have studied for years, most of them without understanding its significance. Cells have partitioned their DNA into two pools for replication and packaging, an early replicating pool and a late replicating pool with a short pause between. It is possible that some cells have invented a third or fourth pool, although I doubt that elaboration is either necessary or desirable. In the first pool, replicated in S_E, are all the genes that the cell will need to use in its differentiating processes and in functional roles. All other genes replicate in S_L, the last half of the S phase, and are maintained in an inactive state. The proteins that sequester genes from transcription are available in S_L. The proteins that open up the genes to transcription are available only in S_E and any sequence replicated in S_E is potentially functional when and if the cellular environment is appropriate. This situation makes the control of the time of replication in the cell cycle crucial, but it probably makes little difference in what part of S_E or S_L a particular gene replicates. It also means that a mechanism to modify replication origins in a way that is stably inherited is a crucial feature of differentiation. Some replicons must be shifted to replicate in S_E, others to S_L.

"Methylation as a part of differentiation may be a late development in evolution and is only extensively exploited in the vertebrates, in which it suppresses those genes in a cluster (replicon) that will not be useful in a particular differentiated cell. For example, in a red [blood] cell precursor all globin genes will be switched to replicate in S_E but only those will be expressed which are demethylated during a subsequent determination step.

"I know that both of these ideas, and particularly the first, will be considered naïve and unsupported by all of the evidence by some critics, but I predict that biologists will still be investigating both these phenomena and acquiring astonishing results when the critics and I have passed from the scene" (DNA Methylation and Cellular Differentiation, 1984).

Taylor at Boone Chromosome Conference, 1975

    Taylor decided he could test the idea that methylation is used to turn off gene function in differentiated tissue. He predicted that, prior to differentiation, cells would have a low level of methylation, and after differentiation, when many genes would have been turned off, the methylation level would be high. Germ-line tissues such as testis and chorion are undifferentiated, whereas mature organs are fully differentiated. Graduate students Karin Sturm and R. Alfred McGraw collected bovine tissues from slaughter houses, isolated satellite DNA, and sequenced it using a new technique that distinguished methylcytosine from cytosine. Sperm and chorion were methylated at few sites compared to thymus, brain, liver, and kidney cells (1981–1990). When the data were analyzed they supported his hypothesis—methylation is correlated with differentiation.

    DNA repair seemed another area where the methylation of DNA might have an important function. Ironically, the very presence of methylation on the base cytosine left DNA vulnerable to a very specific type of damage. Methyl cytosine in a G:C base pair can undergo a chemical conversion that converts it into a G:T mispair. If the mispair is not corrected, subsequent DNA replication produces, in half the cases, the original G:C pair but, in the other half, an A:T, a mutation. Taylor reasoned, first of all, that methylation must have a very important function to be worth even a small risk of this type of mutation and, second, that the mammalian cell must have a very efficient way of recognizing which of the bases in the mispair was correct and should be retained so that mutation could indeed be held to a very low rate. Graduate student Joan Hare took on the project. Together they designed a model test system using SV40. In the early years of molecular biology, SV40 was one of the first viruses of eukaryotes for which complete restriction maps and DNA sequences and gene functions were known and so much genetic information was already available—a perfect model system. This research also marked a period of time in the Taylor lab—the molecular biology period—when the much more manipulable viruses were used as model systems for their much larger host cells.

Indeed, Taylor's instincts were correct; methylation itself, on the strand opposite the damage, was able to direct the cell to correct the mispair so that the original G:C pair was returned in a high percentage of the DNA molecules. When the opposite DNA strand lacked methylation, the ratio of the types was much closer to random. Together Taylor and Hare published the first paper that demonstrated that methylation at specific sites in mammalian DNA serves as a mechanism by which new and old DNA strands are distinguished during replication for the purpose of repair. Hare continued to work with Taylor on this project for several years after she received her Ph.D., and together they looked at more of the patterns of methylation affecting mismatch repair, eventually also discovering factors other than methylation that served as DNA strand markers in mismatch repair.

◇

Origins of DNA Replication

The search for the origin of DNA replication in eukaryotic cells was one of the pressing challenges of molecular biology during the late 1970s and early 1980s. A single site had been identified in the E. coli chromosome where DNA replication always began. Investigators hoped that such sequences could also be found in eukaryotes. Herb Taylor's laboratory was among those that had shown that eukaryotes initiated replication at thousands of sites, and he had further shown that under certain conditions the number of sites could be made to increase severalfold. His lab seemed to have a good chance at identifying these sites. Herb and graduate student Shinichi Watanabe set out to clone DNA replication origins. The restriction enzyme EcoRI had just become available and the methods of cloning just developed. Watanabe used EcoRI to fragment the DNA of Xenopus and clone it into a plasmid of E. coli. After cloning hundreds of these fragments, he injected individual clones into Xenopus eggs, looking for one fragment that had the ability to replicate. One clone replicated far better than the rest. It was sequenced, and by functional definition it was identified as an origin of replication.

Several laboratories identified sequences that appeared to have functional properties of DNA replication origins. Taylor's group had identified one of the first.

Of course, the replication story became more complicated. Further dissection of the sequences did not seem to clarify the picture. Twenty years later, molecular biologists still can't define what makes a DNA sequence an origin of DNA replication. Herb Taylor defined several pieces of the puzzle; more have been supplied since his work ended, but more still wait to be supplied.

Election to the National Academy of Sciences

In 1977, twenty-four years after the groundbreaking work for which he would always be remembered and honored, J. Herbert Taylor was elected to the National Academy of Sciences by his scientific colleagues. It had seemed

Taylor at Florida State University commencement ceremony, 1984

Florida Governor Reuben Askew (fourth from left) and Florida State University President Bernard Sliger (fifth from left) honor new National Academy of Sciences Inductee J. Herbert Taylor (third from left). Also present were three other academy members from FSU: Paul Dirac (far left), Abba Lerner (second from left) and Lloyd Beidler (far right).

a long time coming, but all of his immediate colleagues and laboratory were very proud. "Dr. J. Herbert Taylor, professor of biological sciences, has been elected to the National Academy of Sciences—one of the highest honors that can be accorded an American scientist," announced the Florida State University newsletter (1977). There were at that time only three other academy members at Florida State University, and a total of only ten in the entire state of Florida. He was joining the national body of about 1000 members.

His election was in recognition that "He, over 20 years ago, performed one of the truly classic experiments of modern biology which provided a foundation for molecular genetics. His study showed that chromosomes were replicated in the same manner as predicted for the newly proposed DNA molecule. His experiments directed early attention to the work of Watson and Crick for which they were later awarded the Nobel Prize in medicine. His more recent work has been concerned with the replications and assembly of DNA into chromosomes in cells of higher organisms including man" (Florida State Bulletin, May 1977).

Former Ph.D. student Marcia Applegate Harris recalled:
"I remember how proud we were when he was voted into the National Academy of Sciences, and how he told us that his colleagues

at Columbia had warned him not to leave and go to FSU because if he wasted his research life at that obscure place he'd never get into the academy. He'd just told them that New York was no place to raise kids and left anyway." (letter, 1999)

Reception at the Institute of Molecular Biophysics, Florida State University, honoring Taylor on his naming to the National Academy of Sciences, 1977. [First photo] with Biological Sciences department chair A. Gib DeBusk; [second photo] with daughter, Lucy, and Shirley, reading congratulatory telegrams.

Chapter 6

Living in the Community —
of Science, the University, and Tallahassee

When the Taylors arrived in Tallahassee no Quaker option was available for their children's continued education, and they wanted something different from their own earlier church connections. Herb had attended a Baptist church in his college years and Shirley a Brethren church college. They found a small Unitarian Fellowship that used the Sophia Fahs comparative-religion curriculum. It was a do-it-yourself group of interesting religious liberals, without a minister, and it was growing out of its space in a beautiful but very small chapel. At this time it was the only location at the university where a black speaker was welcome. In the early 1970's, Herb was elected fellowship president and inherited the pressing task of constructing an affordable new home for this small group. Fellow Florida State biologist and Unitarian Dexter Easton remembers.

Dexter Easton: "With characteristic calm and enthusiasm, Herb, negotiating directly with the builder, brought the cost of the building down to a reasonable $80,000, a considerable saving from the original bid of $130,000. To do this he oversaw a fundamental change in the architect's plans.... Herb rejected the original conventional design as being too walled-off from the location's forested world and substituted glass windows for solid sanctuary walls. Just as he had earlier rejected conventional ideas of chromosome structure, with his special insight he brought light into the cell nucleus and into this church ... Herb knew what was important: in biology he realized that to understand the function of the cell nucleus, we ought to know the structure of the chromosomes; in religion, he understood that theological truth should endure the light of science. We are indebted to Herb for his foresight, imagination and dedication in science and in humanism" (reading at memorial gathering and letter, 1999).

J. Herbert Taylor, 1980's *Shirley H. Taylor, 1980's*

By 1970 Shirley had taken on an additional role with the university graduate dean, to survey all departments for the feasibility of forming an environmental studies program. Continuing part-time to manage Herb's lab, she worked independently on environmental policy issues—Florida wilderness, coastal land and water ecosystem protection, offshore oil impacts—and developed and led campaigns for new federal protective legislation. Herb encouraged and supported her efforts as she developed her

Taylor, exploring Florida Everglades mangrove canoe trails, 1984

Leading cytologists, in San Francisco at one of the first meetings of the American Society for Cell Biology; Wolfgang Beerman, J. Herbert Taylor, Joseph G. Gall, and H. G. Callan, 1962.

"second career" in national Sierra Club leadership roles on coastal issues over the next 25 years.

Herb Taylor was one of a group of 16 scientists who formed the American Society for Cell Biology in 1960, and he served as its president in 1969–1970. The organization soon became the most influential in its field; attendance at its annual meetings now exceeds 5000. He was a charter member of the Biophysics Society and worked in editorial positions for a number of scientific journals.

In addition, Taylor had taken on the role of editor of Chromosoma in 1966. Starting in 1984, he shared the tasks of managing editor with Wolfgang Beermann and Wolfgang Hennig, two other cytogeneticists. Taylor remained an editor of Chromosoma for more than 20 years, until his retirement from FSU. Hennig, the current Chromosoma editor, remembered meeting Taylor first at Beermann's institute in Tübingen and later at various scientific meetings and Chromosoma editorial board meetings.

Hennig: "I was always impressed by his clear judgment, his broad scientific expertise and, in his function as an editor, by his responsible handling of manuscripts—not letting unscientific arguments enter into his decisions on acceptance, revision or rejection. I was always impressed with his expertise and excellent advice, Herb Taylor guided and contributed to the development of this journal. All of us who benefit from *Chromosoma* today—readers and authors, editors and publisher—are grateful to him" (*Chromosoma*, 2000).

Taylor at the Institute of Molecular Biophysics, Florida State University, 1977

His calm handling of people and astute scientific judgment were also valued by his FSU colleagues. After serving as Associate Director of the Institute of Molecular Biophysics for ten years, J. Herbert Taylor became Director in 1980 and served until 1985. Mentoring his students, teaching, and administration left him less time for hands-on research, but he continued to write and pursue answers to DNA puzzles.

In 1983 the Florida State University faculty bestowed their highest honor on their colleague J. Herbert Taylor. He was named Robert O. Lawton Distinguished Professor. The candidate must be a tenured professor, have been at university for at least ten years, and have achieved true distinction nationally and/or internationally in his discipline or profession. Although scholarly distinction is the primary qualification, emphasis is placed on the evidence of high-quality teaching, including the direction of graduate research, and service to the university and academic community.

In the spring of Herb Taylor's retirement in 1990, his graduates Joan Hare and Terry Myers arranged a weekend celebration that brought back dozens of his former students from around the country. In addition to fellowship and food, of course a canoe trip was organized.

Herbert Taylor retired with rank of Professor Emeritus in 1990 at the age of 74. He felt, just as Henry Thoreau had written, that "if one advances confidently in the direction of his dreams, and endeavors to live the life he has imagined, he will meet with a success unexpected in common hours."

Herb Taylor with his Ph.D. graduates, Jesse Sisken and Terry Myers.

Taylor continued to write and travel and enjoy life even as he battled prostate cancer for a dozen years, until his death in 1998.

Lynne Wall (postdoctoral associate): "His curiosity was infectious and his incisive analytic approach combined with gentle good humor rubbed off on us all. Being in Taylor's lab meant doing all things with good grace and with a zest for life—whether that was picking wild berries, listening to music, visiting the wilder parts of Florida, or speaking up at lab meetings" (letter, 1999).

Bill Haut (Columbia graduate): "Herb was a patient and nurturing man ... He nurtured his wife and children, he nurtured his students, and he constantly nurtured ideas—in himself and others. And he was generous in sharing those ideas" (letter, 2003).

Writing in My favorite cell with large chromosomes *in 1991, Taylor concluded:* "But my hopes and dreams of solving some of the problems of meiosis and the mechanism of crossing over that began as a graduate student have led me on an interesting odyssey where I met many helpful and friendly people. They did not always agree with my choices or ideas, but few were discouraging. Many were willing to listen to what I had to say or to read what I wrote. To paraphrase a fellow Oklahoman,

Will Rogers, I have never known a man or woman that I did not like. There must be some who are thoroughly detestable, but perhaps I moved on before I learned to know them."

Roaming the Earth

Herb and Shirley's children, Lucy, Michael, and Lynne, Thousand Oaks, California, 1998.

By 1980, the three Taylor children had grown up and were finding their own places in the world. Lynne earned a master of arts in early childhood education and married an Australian; they have two children. She is now director of Boopa Werem School for preschool aboriginal children in Cairns, Australia. Lucy is an artist in various media and specializes in ceramics, teaching a four-year high-school ceramics program in Agoura Hills, California. Michael Taylor, with a master of science in optical engineering, works in the semiconductor industry. In 2003 Shirley joined him, his wife, and their two boys to resettle in Williamstown, Massachusetts.

It was after their three children were grown that Herb and Shirley traveled to wilderness and exotic places of which they had always dreamed. They traveled the Nile north from the Aswan Dam by ship, stopping for dusty walks to Egyptian tombs and temples. Taken to

Shirley Taylor at home in Los Gatos, California, 2000

see Egypt's population-control program, they were the first Americans ever seen by these Delta villagers. In Alexandria they were hosted by former student Effat Badr, then a professor at that university. They shared the year's travel experiences with friends through letters each December. Looking back 50 years to the November when he completed his Ph.D. degree requirements and began Army service, Taylor wrote to his former students:

Herb: "We have continued to roam the earth looking for the wildest and most primitive places we can find. Beginning nearly 40 years ago in trips from Manhattan to Colorado in a used Volkswagen camper, we have searched all 50 states, most provinces of Canada, Mexico, and Central America and parts of Thailand and Bali. High points include camping in tents across Tanzania's Serengeti Plains with fires burning at night to keep the roaring lions away; our trip to India and Katmandu and watching tigers in Nepal; seeing Australia's Outback from Cairns to Darwin and our camping trip to the primitive home of Aborigines in Arnhem Land, then by bus the length of Australia through Alice Springs to Adelaide and Sydney; and finally an Elderhostel trip to the primitive parts of Fiji, Tonga, and Western Samoa" (letter, 1993).

Herb on a camping trip across the Serengeti, Tanzania, 1985

Herb and Shirley Taylor in Thailand, 1991

Shirley and Herb Taylor receiving a gift, India, 1983

Herb and Shirley Taylor and former student Dr. Vinod Shah as guests at Varanasi, India, 1983

Taylor with former student Dr. Vinod Shah, Dean of the School of Sciences, Gujarat University, Ahmedabad, India, 1983

On their trip into Ruanda to visit the mountain gorillas, they were impressed with the overwhelming problems of overpopulation. Already an advocate of family planning, Herb became an active supporter of Planned Parenthood and lobbied for U.S. aid to family clinics abroad.

Herb: "We did go to Belize (formerly British Honduras) this spring to visit Mayan ruins there and Tikal in Guatemala and to see subtropical vegetation and snorkel over beautiful coral reefs. During August and September we went to the International Genetics Congress in Birmingham, England, where we saw former students Effat Badr and Deepesh De, both with active research and administrative careers. We took side trips to London and Stratford-on-Avon to see a play at the Shakespeare Theater. After the congress we spent a week in the mountains and coastal regions of northern Wales, a week on one of the narrow antique canal boats cruising some of England's old canals, and finally a week exploring in the south of Ireland before returning via Shannon Airport. There are at least two other places I have seen by TV that I would like to visit, the primitive northwest coast of Australia and the Okavanga delta in Angola" (letter, 1993).

Herb on a canal boat in Wales, 1993

And they did manage to reach these two distant places:

Shirley: "After spending June revisiting many of our friends, Oklahoma family and favorite spots in Colorado, we were off to Botswana for a two week camping safari. It was a hard trip, 900 km, in a 4-wheel drive vehicle on rough track or deep sand. We traveled across the Kalahari Desert spotting springbok, ostriches, and oryx. On the Okavanga delta we camped yards away from a hippo 'pool' (lake), and saw herds of elephants destroy forests (they push over trees to eat just some of the leaves), and roll over and over to 'mud' themselves. We saw herds of zebra, a lion pride stalking a herd of vicious Cape buffalo (lions were finally defeated in head to head combat). Victoria Falls Hotel in Zimbabwe (shades of British Raj in India!) was a great, and welcome, culture shock to us, covered as we were with gray desert dust" (letter, 1995).

Shirley: "Before visiting our daughter, Lynne, we flew to Darwin and then by Alligator Airways' tiny pontoon plane to King George River on the wilderness Kimberley coast of northwest Australia for a weeklong cruise. We crawled across to board the Opal Shell, a sailboat crewed by a couple for just three passengers. Gourmet meals on deck under shade from a hot sun, forward cabin bunks, great fishing, virgin beaches, cruising deep rivers with high walls of weathered red stone—and total wilderness. We saw only two small boats and one hermit (but many crocodiles and birds) the entire week during which we explored three deep rivers, beaches, uplands, and islands. Dream fulfilled!" (letter, 1996).

Herb and Shirley Taylor fishing in the noon sun, Kimberley wilderness river, north Western Australia, 1995.

Shirley: "We surprised ourselves with an unexpected trip, by flying west to Siberia, then traveling two days by trans-Siberian Railway to visit Irkutsk and take a week's cruise on Lake Baikal with a group of Russian scientists. It was a different and interesting week, new ideas, new fish, new foods. We explored meadows and islands with varied flora and rare species, and carefully hid ourselves to watch the only fresh water seals in the world. We visited an isolated fishing village, mired in desperate post-Communism poverty. We learned much from the Russian scientists about their world" (letter, 1997).

Herb with the "Brown Bear" camper he custom built, 1970's

Herb with another camper he custom built, "Polar Bear," 1990's.

In 1997 Herb and Shirley flew in for a return visit to a remote wilderness lodge on Lake Wabatongushi in northern Ontario. Flying in the small plane gave them a better view of the many small lakes and forest types that cover so much of that world. Steering a cedar-strip motorboat under sun, and sometimes wind or rain, they explored and fished mile after mile of this very long, narrow lake. Now and then they caught sight of moose, bear, and eagle and once had to push hard to avoid wind-stranding on a rocky shoreline. They managed to reach camp, with their fish, before sunset.

That day was like their lives, with many high moments and a few low, but altogether satisfying and with surprises along the way. As that brilliant sunset darkened to evening, the loons became active on Wabatongushi waters, trilling their silvery notes on through the night.

    All citations of Herbert Taylor's work may be found listed in the curriculum vitae at the end of the book. For citations within quoted materials, the reader is referred to the references within the original citation.

    All footnotes are from Webster's *Seventh New Collegiate Dictionary;* Facts on File Dictionary of BioTechnology and Genetic Engineering.

Part II
Tributes

A Memorial Gathering to remember J. Herbert Taylor was held at the Unitarian Universalist Church of Tallahassee in Tallahassee, Florida, February 6, 1999. Family, friends, and colleagues, as well as scientists whom he had trained spoke. Distant friends and colleagues also contributed their remembrances as letters over the following weeks and months.

Memories of Herb Taylor as Father, Brother, Brother-in-Law

Even when the candle is out, the wick glows for long, the fragrance lasts forever

– Deepesh De

Michael Taylor

A son's memories of his father

Herb Taylor was my father. His passing has left a hole in my world and in the lives of those who loved him, but he remains in our hearts and our minds, sometimes in distinct memories, but often tucked away in the quiet corners of ourselves that we don't even notice. Who he was has become a part of me and my children in a thousand ways, many that I will never even know.

My father truly was a self-made man. As a child I remember him as a man of towering intellect, a seemingly endless source of information and answers. He was a caring father, but the most important lessons he left me with were sometimes not in what he did, but in what he did not do.

When I was about 8 years old—a cub scout—we had a contest to race rockets. They were powered by rubber bands and propellers and suspended from wires across the school cafeteria. Each kid got a kit of parts to make their own rocket. It was a complicated job, carving the wood for the nosecone, painting, assembling all the parts. Frustrated with the job, I asked my Dad to help—but he said he wouldn't do it for me. He would answer my questions, but I had to do the job myself. Bit by bit I got it together, although it was rather crooked and the paint a little messy. I went to my friend's house to check on his progress and was amazed to see his rocket with gleaming enamel paint, beautiful pinstripes and decals—until I realized that he hadn't done it at all—his father had built it for him.

On race day I felt embarrassed to bring out my rocket. It looked like such an ugly duckling compared to the gleaming rockets of my friends—although most had really been done by their fathers. I wound the propeller extra tight, hoping to overcome my rocket's handicaps, but the rubber band broke and tore the rocket in half. I went home dejected, while my friend's "father built" rocket took first prize. A father now myself, I have told this story to my boys many times. In that version, the next morning

I awoke, realizing that I had in fact won the most important prize of all, gaining the knowledge and skills from working through this experience on my own. My friend by contrast had only a useless trophy on his shelf. In fact, it was not till many years later that the importance of this and many other lessons became clear to me. I know now how hard it is to stand back while a child struggles, and to let them solve their own problems, with support but not intervention. It has been one of my father's most important and lasting gifts to me.

My father had a quiet confidence and self-taught ability to do so many things, often with an unconventional approach. A wonderful cook, he never seemed to be able to make the same thing twice, generally avoiding recipes. He was an avid nonreader of instructions, preferring instead to discover for himself the operation and nuances of the latest VCR or camera. He was an avid photographer and skilled in the darkroom. He loved to fish. He told me once that the challenge was to outsmart the fish. And occasionally he did. He was a carpenter, plumber and electrician, mason, roofer, lumberjack, gardener. Working with my mother's father, Dr. Roy Hoover, he became a skilled woodworker, building a household full of fine furniture. There was virtually nothing that needed doing that he wouldn't at least try to do himself.

Growing up in the shadow of a man who had achieved such success in his professional life was sometimes very intimidating to me. I will always be grateful for his loving support and for helping me find my own way. I cherish his memory, all those bits and pieces that are tucked away inside me, and all those whose lives he touched.

Lynne Taylor Ireland

His daughter from "down under" remembers:

My father nurtured my interest in the hows and whys of the world from the time I was a small child. He was always ready to listen to and answer questions and discuss ideas seriously. He shared his enthusiasm for knowledge and discovery. One memory that stands out is being totally intrigued as he explained the idea of an infinite universe when I was maybe 7 or 8. From him I learned also a healthy scepticism, to examine ideas critically while keeping an open mind. Later, as an adult, I always looked forward to hearing his ideas and opinions on many issues—world events, economics, and politics as well as science. He was well informed and thought through issues carefully. He was an enthusiastic believer in Do It Yourself and enjoyed teasing apart the problem of how to do something or why something wouldn't work properly—from fixing an electric razor to major building projects. Following the example he set has allowed me to experience the joy and satisfaction that comes from thinking through and solving problems. My father was a keen observer not just in matters of science but of people and places as well. He had a good sense of humour and could tell a good story. I remember listening to entertaining tales of his boyhood in Oklahoma as well as of people he had met or places he had visited.

I feel lucky that I grew up in a family that encouraged such an interest in understanding the world, in asking questions and exploring ideas, because these "habits of thought" are so important in any field of endeavour and in living a rewarding and satisfying life.

◇

Lucy Taylor

An artist daughter "learned by osmosis"

It is often said that children learn a lot from their parents by osmosis. While my father was researching the mysteries of science, I was a child looking at the world and asking my own questions. Why is the sky blue? Why is the grass green? What makes a rainbow? It was fun to ask, especially when I knew that I could find out. I learned to wonder, to be interested and amazed, to look at the world around me and ask why?

My father provided answers to any and all my questions with a scientific explanation appropriate to my level of understanding. The questions developed as I did. As I grew older, the queries evolved as did the answers. How does the immune system work? What is the difference between a virus and bacteria? In the way of children, I took all this for granted. Wasn't everyone's father able to explain anything to them? Didn't all parents discuss the botanical names of each plant they saw while hiking in a forest? Didn't everyone look at the world around them and wonder how and why things happen?

As an adult, I have come to understand that what I learned from my parents was special. I have learned to think about things in unique ways. The questioning and the wonder have become a part of me. I always want to know how and why things work? These days, I have to go various places for my answers. I no longer have one resource easily available. But several years back even as the miles separated us, my father was happy to provide information at my request. I was watching a science fiction movie that depicted the detonation of a nuclear warhead in a space station, complete with explosions, fire and shock waves. I started wondering, "Could that really happen, fire and sound in the vacuum of space?" I needed to know. The easiest way to find out was to ask my dad. I sent him an e-mail, outlining my questions. In true form, he replied quickly with a specific response. He answered all my questions, as well as taking the time to explain nuclear fission and fusion, the difference, and the possibilities.

What is the legacy from my father? I learned to ask questions, to inquire, to see the world around me with wonder. I can't imagine living any other way. I always need to know why, and it is always important to find out. I need to know how things work. I naturally look beneath the surface. This is my inheritance. It is part of who I am. I am the artist and the mystic in our family. I have applied this curiosity to numerous subjects. My lines of inquiry have evolved along many paths. But because of the nature of my upbringing, I have been inspired to ask some of life's deepest questions, and I am finding answers.

◇

Tom Dodson Taylor

Herb's proclamation: his brother's story of how Herb left the farm and became a scientist

Herb Taylor and his brother Tom Dodson (T.D.) Taylor were operating a "binder," the large farm machine for cutting and binding oats or other grains. Here's how T.D. recalls that day:

We were binding a late, overgrown field of oats in a river bottom in southeastern Oklahoma near the Red River. It was late August, humidity near 100%, temperature around 100°F. A wet summer had caused the oats to grow shoulder high, and the binder would stall when we hit spots of wet soil. The binder was a heavy load for the four mules to pull, especially in the tall oats and in wet soil. The machinery of the binder was run off a "bull wheel," a large wheel whose chain gear hooked to all the binder's operating equipment. When we hit a soft wet spot the bull wheel would slip, and all the moving canvas and cutting blade would jam up with hay. We would then have to pull all the jammed up hay out of the equipment, and finally turn the gears with a hand crank to clear hay out completely. It took both of us to get the crank started and then several minutes of turning. Every fifteen minutes we were forced to repeat these operations.

When we had stopped for perhaps the tenth time that afternoon we were standing back to catch our breath and cool off a bit while we studied the sky, wondering if we might get an afternoon thunder storm. I noticed that Herb had stepped off a bit, looking over the whole situation in a careful study. I thought maybe he was considering whether we should pull out to let the ground dry more. Instead, he looked square at me and proclaimed, "I don't know about you, T.D., but I am not going to spend any more of my life doing this!" We pulled out of that field, secured the equipment, unhooked the mules, and headed for the barn just ahead of a thunderstorm.

About a week later Herb was off to the CCC Camp, where he spent the next year. He enrolled in college the next fall, and sure enough—he never returned to that river bottom oat field or any part of farm life again. He found a much more suitable calling.

(Read by Ralph Cook at the Memorial Gathering)

June Hoover Foster

Admiration from a sister-in-law

Dr. James Herbert Taylor was my brother-in-law. He was a man whom I thoroughly respected and of whom I felt in awe at times for his outstanding accomplishments in the field of molecular science. Because music and the fine arts were my field of interest, my knowledge of the sciences fell somewhat by the wayside, and I believe Herb realized this. He never tried to embarrass me by trying to discuss or explain his laboratory endeavors to me directly. This I truly appreciated.

Besides being an expert in the lab, Herb had many other fields of interest in his life. Did you know he was an excellent cook? He could turn out some tasty dishes and serve a great meal when the occasion demanded it. Gourmet meals I would call them. Of course, Shirley would always add her feminine touch, too.

Did you know Herb was an expert carpenter and fine cabinetmaker? He completely outfitted two different vans, transforming them into living quarters for two people, providing sleeping area with toilet facilities, a dining table, and kitchen accommodations plus storage spaces. The first outfitted van was referred to as "The Brown Bear" and the second as "The Polar Bear," their colors being brown and white, respectively. These vans facilitated Herb's and Shirley's many trips into all parts of the United States and Canada, using not the well traveled highways but the less charted roads and byways to see and enjoy the great outdoors and nature's marvelous store of plants and animals.

Herb also found time to renovate part of their home basement into an apartment for Dr. R. M. Hoover, his father-in-law, so that it would completely accommodate wheelchair living during Dr. Hoover's waning years. That was no small achievement.

Two other special interests of my brother-in-law that clearly raised him into the realm of a nature lover were his developed skills with the canoe and with fishing. Moments in the canoe and waiting for the fish to bite gave him time to think, to ponder, and to come up with some answers from time to time, so I heard him say.

As I saw Herb, he was a loving husband and a caring father who enjoyed his children and later his grandchildren, watching them grow and mature and always showing pride in their achievements. He was a concerned father when things went wrong but ready to laugh when things went right or were amusing. Herb was the first to laugh at himself when he made a wrong calculation or an out-and-out mistake.

Herb was a quiet person, a deep thinker, and often sat in silence and listened to conversations without entering into the discussion unless called upon. Perhaps he would be lost in his own thoughts and processes. Of this I was never certain, but it was this special quality about Herb that I truly admired—no lost words with him! He was a highly respected person, an admirable person, and I shall miss him.

◇

Memories from the Community

*The consequences of one's life have
enduring echoes in the lives of others.
Herb created many echoes and we shall miss him.*
— *Martin Roeder*

Michael Kasha
Institute of Molecular Biophysics
Florida State University
Tallahassee, Florida

His beginnings at Florida State University

Herbert Taylor—professor, naturalist, craftsman, friend—came to the Florida State University in 1964 from Columbia University, where he had served since 1951. Dr. Taylor made notable contributions to the science of molecular genetics, of which he was one of the original developers. He proved in work done as a Columbia University professor, in collaboration with colleagues at the Brookhaven National Laboratory, that DNA in higher organisms preserves its continuity as unbroken molecular chains during replication. This was one of the first experiments to confirm the Watson-Crick model of replication. The replication experiments involved using a tritium-labeled DNA base (thymidine). The radioactive isotope of hydrogen permitted autoradiographic tracing of the mechanisms. Some "crossovers" from one chromosome to the other were observed, leading to the concept of the "semiconservative" replication of DNA.

The Institute of Molecular Biophysics had been established in 1960, and after a nationwide search Herbert Taylor was selected for the institute by the Biological Science faculty as the first professor of molecular biology. In 1983 he was appointed a Lawton Distinguished Professor, a position he held until becoming Professor Emeritus in 1990. I had been the Institute director from 1960 to 1980, resigning at that time to establish some directorship rotation. Herb Taylor had served as associate director for biological sciences from 1970 through 1979 and was appointed Director to serve from 1980 through 1985. His deep knowledge of biological sciences and his integrity and generosity as a human being made him the natural choice.

Dr. Taylor maintained an active research program throughout his academic career. He edited a notable series of three volumes of collected papers in *Molecular Genetics,* published between 1963 and 1979 and a reprint volume in *Perspectives in Molecular Biology* in 1965, all under the Academic Press imprint, as well as *DNA Methylation and Cellular Differentiation* in 1984 in the Cell Biology Monographs series published by Springer-Verlag. He was also a member of an organizing group that launched the influential American Society for Cell Biology and served as president in 1969-70.

Herb Taylor's work on DNA strand lineal topological integrity was of Nobel Prize caliber. He was elected to the National Academy of Sciences in 1977 in recognition of the value and originality of his research in molecular genetics.

◇

Kurt Hofer
Institute of Molecular Biophysics
Florida State University
Tallahassee, Florida

The young colleague he befriended when he arrived at FSU

I have known Herb Taylor for nearly 30 years. He was a giant in my field of research, but more important, he was also a warm and generous friend. He was always willing to help and to share his ideas. Of course, I was not the only one who benefited from Herb's sterling qualities; he was highly regarded by all of his many colleagues and friends. A few years ago I came across an article entitled "On tritium, DNA, and serendipity" written by Bob Painter, one of the most influential radiation scientists of our time. In this article Bob Painter describes his association with two legends of modern science, Herb Taylor and Pete Hughes.

"Early in the summer of 1956 Pete Hughes informed me that some fellow by the name of Herb Taylor was visiting Brookhaven and wanted to use 3H-thymidine for a tracer experiment he had devised. Pete Hughes had already synthesized small quantities of the compound and agreed to supply it to Taylor. Later that summer, Pete called me again and said that this fellow Taylor was very excited and wanted Pete to see his results. Pete took me along, and I remember looking through the microscope at autoradiographs and wondering what all the fuss was about. It took a couple of explanations by Taylor before I realized that this was actually a demonstration of semiconservative replication of DNA and that I had fortuitously stumbled in on one of the most important developments of modern biology."

As it turned out, this discovery was not only a milestone in the history of biology, it was also a crucial turning point in Bob Painter's career because at that meeting Herb suggested an idea that determined the future of Bob's research. It was a piece of sheer good luck, but as Bob Painter said in the article, "I'd rather be lucky than good, any day."

And that is precisely how I feel about Herb's role in my life. I too was lucky to have met him. In my case, Herb helped me to set up my laboratory and to start my teaching program. We jointly developed and taught a course in cell biology for more than 10 years. In my graduate program, Herb was an advisor to every one of my first 15 graduate students. In my research, Herb and his wife Shirley have helped in countless ways, both little and big. Shirley taught my wife Ridy the mysteries of tissue culture, and Herb spent innumerable hours of his precious time in helping me to unravel some of the problems I encountered in research.

Clearly, Herb was not just an outstanding scientist; he was one of those rare individuals who make a difference in the lives of others. As we say good-bye to Herb, we know that we have suffered a great loss, but we also remember how much we have gained. Herb Taylor gave us the greatest gift anybody can give, because he gave of himself. So today we thank you, Herb, for what you did. We will miss you a lot, but we are lucky indeed to have known you.

◇

Eleanor F. Moore
Tallahassee, Florida

Memories from an enduring friend and fellow adventurer

I have many happy memories of adventures with you and Herb while going on canoe trips. Of course the near disasters stand out in my memory—the time you and Herb were in your canoe ahead of your father and me. The stream was high and fast. Dr. Hoover and I were pushed up against a downed tree. I ignorantly pushed against the tree and over we went. The water was cold! Our lunch, jackets, and equipment floated out of the canoe. We called for help and you turned around and paddled back up stream, picking up the contents of our canoe as you came. I don't remember how we got out of the water. I do remember that you had dry shirts and jackets. Those dry clothes allowed us to continue our trip.

You and Herb were calm throughout our canoe trips, the building of our church, and the research done for the Sierra Club. Thank you.

Love,
Eleanor (F. Moore)

Lawrence G. Abele
Provost,
Florida State University
Tallahassee, Florida

Dear Shirley,

Thank you for your kind message. Linda and I are with our grandchildren right now and are loving it.

My favorite recollection of Herb was the privilege of attending a talk he gave to undergraduates on his career. Herb began with how his love of science and the help of teachers pushed him on to attend college. He then talked about his work on DNA and it was so exciting that I felt like I was there participating in some of the most important work on genetics of the century.

It is wonderful that we are able to work together on an FSU memorial to Herb.

Back to worshiping the children, as they wish to be fed and you know that we take care of their every wish.

Larry Abele
(Biological Science professor and colleague, FSU)

Bob Hopkins
Sierra Vista, Arizona

To my friend:
Dear Herb,

I write this in total enduring and loving friendship. How wonderful it was to discuss with you—science, gardening, life and loving … not to mention one's passions. Yes, the sparkle in your eyes, that wonderful laugh of satisfaction, the wonder of the stars, and this world of endless explorations will always remind me of you. I did like your jewelry, the enamel pieces and the soul of their inspiration. Thank you for sharing so much with us in that old Civil War Pennsylvania farm house. I will always treasure you as a friend, a guest, a loved one.

Let's walk together again in your Tallahassee gardens. Let's walk together again in my Pennsylvania gardens—remember the iris, peonies, daylilies, azaleas, rhododendrons, and all of the special secret places.

Now, I welcome you to join me in my new Arizona desert gardens. The plants bite, but the flowers are so beautiful. How about making some prickly pear jam with me? I'll be expecting you and Shirley. It's a date for sure.

In love and admiration,
Bob Hopkins
(Retired high school science teacher, partner of former graduate student Bill Haut—his father was painter and Chair, Art Department, Hofstra University, Long Island)

Joseph Travis
Department of Biological Science
Florida State University
Tallahassee, Florida

Dear Joan,

Thanks for distributing the remembrance of Herb Taylor [from the American Society for Cell Biology Newsletter]. I remember meeting him when I was a new assistant professor and being impressed by his gentleness, generosity of spirit, and absolute lack of pretension. Would that all of us could be more like him.

Joe Travis
(Colleague and former chair, FSU Department of Biological Science)

◇

Hank Bass
Department of Biological Science
Florida State University
Tallahassee, Florida

Mrs. Taylor,

I teach undergrad genetics in the Biology Department here. I made this handout (In memoriam: J. Herbert Taylor, from *Chromosoma*)—thought you would like to know. Our textbook (Klug et al., 6th edition) has two full pages on Herb's 1957 paper, describing how important it was. I thought it would be good for the students to know he was at FSU in their majors department.

I didn't know Herb personally, but have known of him and heard many wonderful things about him, as person and scientist. I understand there's to be an endowed chair in his name. I think that's wonderful and appreciate your continued involvement in that.

I'm a third year assistant professor in biology. I do research in cell and molecular biology of meiosis in maize.

Please accept my apologies for this unexpected letter, and I am sorry for your loss. I hope you are well and have a nice spring and summer.

Sincerely,
Hank Bass

◇

Harold Van Wart
Biochemistry and Molecular Biology,
Roche Pharmaceuticals
Palo Alto, California

It was so nice to hear from you—I'm only sorry that the occasion was so sad. I appreciate your informing me of Herb's death since I am so out of contact with events at FSU. He was a kind and gentle man with great scientific insights. I did not know anyone who didn't like him. It would make me happy if this quote were attributed to me at his Memorial Service. I actually never worked with him, or even served on the same faculty (though we worked in the same building). I liked and respected him.

Hal Van Wart
(FSU Chemistry professor, 1976–1991)

◇

Martin Roeder
Department of Biological Science
Florida State University
Tallahassee, Florida

Dear Shirley,

Rae and I were so sorry to hear about Herb's passing. Of all my colleagues he was the gentlest and most helpful. Your sorrow and loss cannot compare with what the rest of us feel, yet we hope that you will accept our deep condolences. As with the famous onion root tip [fava bean] demonstration, each generation passes on part of itself and the consequences of one's life have enduring echoes in the lives of others.

Herb created many echoes and we shall miss him.

With deep sorrow,
Martin Roeder

◇

Jack Winchester
Department of Oceanography,
Florida State University
Tallahassee, Florida

FSU colleague in Oceanography and fellow outdoor adventurer

Herb had a real spirit of adventure in the outdoors, a true biologist and keen observer. Of course we will always remember his steady support with Shirley of protecting the environment for the good of us all. But he also understood the life of nature, and we learned a lot from him. We will never forget the time, while canoeing across the Okefenokee Swamp, when we passed what looked like a dead alligator half submerged in the water. Most of us exclaimed how sad this was to see such a tragedy in so remote a place. But Herb suspected otherwise and jabbed the beast with his paddle. It turned out to be very much alive, immediately lurched up and swam quickly away—luckily not toward us!

(Spoken at Memorial gathering)

Dexter Easton
Department of Biological Science,
Florida State University
Tallahassee, Florida

An FSU colleague remembers Herb for his community spirit

I was acquainted with Herb Taylor in two situations. First, we were both members of the Florida State University Department of Biological Science, which he joined in 1964. Second, we were both members of the Unitarian Church.

It is not my place to honor Herb's accomplishments in genetics, especially since others, infinitely more qualified than I, have already done so. Rather, my intention is to remind you of at least one of his nonscientific contributions to human welfare. More explicitly, the fact that we can now assemble in this pleasant meeting place is owing to Herb's dedicated efforts, when he was president of this Unitarian-Universalist congregation three decades ago. I am qualified to tell you this, because I directly followed Herb as president.

Herb and Shirley joined the church soon after they arrived in Tallahassee in 1964. At that time, a decade after its founding, the Unitarian Fellowship was housed in a small building on Wildwood Drive, at that time near the periphery of the FSU campus. It was fortunate for this small group of religious liberals that Herb and Shirley arrived when they did. The group recognized a good opportunity to get things done, and soon elected Herb president. His most pressing task concerned the construction of an affordable new home for the growing but still small congregation. With characteristic calm enthusiasm, Herb, negotiating directly with the builder, brought the cost of the building down to a reasonable $80,000, a considerable saving from the original bid of $130,000. To do so, Herb oversaw a fundamental change in the architect's plans so that we have the result in which you now find yourselves.

Whether or not you have been here before, we hope you are pleased by the feeling of openness and clarity that this building provides. Herb rejected the original conventional design as being too walled-off from the world, just as he had earlier rejected conventional ideas of chromosome structure. His special insight brought light into the cell nucleus and into this church.

Ground was broken for the construction of this building in February 1968. The building was dedicated and Herb turned the keys over to me, his successor, in October of the same year. He then took off for another presidency—that of the American Society for Cell Biology. This building is the culmination of a dream of a few people, especially Herb Taylor.

There are other things in this building that we owe to Herb in one way or another. When Dr. Roy Hoover, Shirley Taylor's father, retired from his profession as a bone surgeon, he converted his skills to working with wood and crafted the lectern, the chandeliers, and the impressive wall hanging. Thus this building will always remind us of Herb Taylor and the Taylor family.

One final thought: Herb knew what was important: in biology, he realized that to understand the function of the cell nucleus, we ought to know the structure of the chromosomes; in religion, he understood that theological truths should endure the light of science. I like to think he saw this Unitarian-Universalist church as symbolic of that illumination. We are all indebted to Herb for his foresight, imagination and dedication in science and in humanism.

Myrna Hurst
Tallahassee, Florida

Dear Shirley,

During the year of 1973, I was faced with the problem of transportation for our daughter when she entered high school. I thought working half time would solve the problem. I was working as Supervisor of the Clerical Office in the Department of Biological Science. I had let it be known that I was interested in a half-time position. Mrs. Rogers (who worked with Dr. Gaffron) came by my office in Conradi and let me know that Dr. Taylor was going to have a position open soon. Also, Dr. Helen Crouse "walked me down the hall" and shared the same information. Soon Dr. Shirley Taylor came by and talked to me briefly, and I made arrangements for an interview. My only previous contact with Dr. Taylor was when he came by to "turn in grades."

This was the answer to my problem—I was elated. I went to see Dr. Taylor, telling him I would probably only want to work half time until my daughter graduated from high school. On September 7, 1973, I went to work for "The Taylors." This was the beginning of a most pleasant and enjoyable 11+ years. I did work for "The Taylors"—they were a team. Dr. Shirley and I shared a small (cozy) office for several years. While we worked, we also "cooked" and "traveled."

It was wonderful working for Dr. Taylor. (I could read his writing!!) I never know of a time when he was so busy that it seemed to bother him to be interrupted to come answer the phone. He was a very patient and kind person, with a smile. (I once asked Mrs. Taylor if he ever disciplined his children when they were growing up! She assured me he did.) In all the years I worked with the Taylors, I was never treated or spoken to in an unkind manner. They never let it be known if anything I did (or said) displeased them. They were great travelers, and they always brought back some "treasure" for me from other cities and countries, which I continue to enjoy today. Many times little notes of appreciation were attached to these gifts, which I still have.

I was enjoying working half time, but as the years passed and I realized I needed to plan for my future retirement, I decided to go back to work full time. I was fortunate to be able to take on the additional duties of the new job and remain in Dr. Taylor's office as his secretary. In 1985, Dr. Taylor stepped aside as Director of the Institute, and since I was the Director's Secretary, I had to relocate with the change. However, I continued to help him with the *Chromosoma* work as long as he needed me.

I can't write this for just Dr. Taylor, as I said they were a "team" and I have to address this to Dr. Shirley as well. To the Taylor family, let me say "Thank You" for my being able to spend 11+ years of my life in such a pleasant working atmosphere and with such wonderful people. He was—and she is—Very Special. This world has lost a gem. I'll always remember his kind, sweet smile and easy-going manner. I wish you well as you face the future.

With sympathy, love, and appreciation,
Myrna
Secretary 1973–1985
(Spoken at Memorial Gathering)

Memories of Herb Taylor as Scientific Colleague

I think that Shirley and his children can take comfort in the knowledge that they spent their lives in the company of a man who, although self-effacing, was a giant in biological sciences.

Sheldon Wolff

Bruce Alberts
President, National Academy of Sciences
Washington, D.C.

Dear Dr. Taylor:

I am writing to extend the sympathy of the Academy membership on the death of your husband. Dr. Taylor was highly regarded by all those whose lives he touched, and although we find solace in the rich legacy of his achievements, his loss is keenly felt by his associates and friends throughout the scientific community.

With my sincere sympathy and warmest personal regards,

Yours sincerely,
Bruce Alberts

Susan Gerbi
Chair Department of Molecular Biology, Cell Biology, Biochemistry
Division of Biology & Medicine, Brown University
Providence, Rhode Island

Dear Joan,

I was so sorry to learn from you about Herb Taylor's death. Have you informed Helen Crouse (his long time research associate)?

The following recollections might be too late for your deadline, but here they are:

I became fascinated with Herb Taylor's work as a high school senior when I had to write a term paper and chose to do it on chromosome structure, and reviewed Taylor's model discussing whether the DNA was continuous or had protein linkers. I very much wanted to study with him, which was a major impetus for me to attend Barnard College. Imagine my dismay in my sophomore year when I learned he would soon leave Columbia University for Florida! I immediately signed up for his molecular genetics class, and had to talk my way into the course, as I had not yet taken organic chemistry, which was a prerequisite. Happily, he allowed me to take this very stimulating course, and despite my lack of background I managed to get an A.

Subsequently, my own research led me to DNA replication (a question I wanted to study for my Ph.D. at Yale, but the methodology was not yet available to answer questions we address today), and I owe my foundation in this to Herbert Taylor. He was a very kind man and a great scientist, and will be missed by many.

Sincerely,
Susan Gerbi
(Past president American Society of Cell Biology)

Sheldon Wolff
Vice-Chairman and Chief of Research
Radiation Effects Research Foundation, Hiroshima, Japan

Dear Joan,

I was saddened to hear that Herb had died. I remember first meeting him almost 50 years ago when I was a graduate student and he came to the Harvard's Bussey Institution to meet my mentor Karl Sax. Ever since that time I followed Herb's work with great admiration. He was a true pioneer who changed the way many things about chromosomes were presented in textbooks.

Several of his earlier studies were seminal works that influenced the development of cytogenetics. He was the first to use tritiated thymidine in autoradiographic studies of chromosome replication and segregation. Back in those days before the advent of what we now call molecular biology, these autoradiographic techniques provided us with the tool par excellence for mechanistic studies. In fact his study showing that DNA segregated and thus replicated in chromosomes semi-conservatively predated the Meselson Stahl experiment. They were at Cal Tech and were fully aware of Herb's experiments. Also in the early history of chromosome labeling, long before we had heard of chromosome painting with molecular biological probes, and of the term chromosome domains in interphase, Herb had shown by labeling the late replicating chromosome with tritiated thymidine that the label was not distributed throughout the interphase nucleus, as common wisdom held.

We in cytogenetics fully appreciated this finding, which was later "rediscovered" with the newer techniques. In other experiments he showed that sister chromatid exchange, which was a quite esoteric and supposedly minor phenomenon, actually was a common event in mitosis. His analysis of twin versus single exchanges, which allowed us to determine whether the exchange occurred during the first or second post labeling round of replication, was brilliant.

Herb's experiments basically were not only innovative, they were simple and elegant, giving witness to the idea that there indeed is elegance in simplicity.

I think that Shirley and his children can take comfort in the knowledge that they spent their lives in the company of a man, who although self-effacing, was a giant in biological sciences.

Shelly Wolff

Bruce Nicklas
Chair, Department of Cell Biology
Duke University
Durham, North Carolina

Dear Dr. Hare,

Thank you so much for letting me know about Herb. I have no specific stories to tell, just a scientific lifetime of admiring Herb as a scientist and as a person.

My warmest sympathy to Shirley and to you all.

Bruce Nicklas
(Past president, American Society for Cell Biology)

Oscar Miller
Department of Biology
University of Virginia
Charlottesville, Virginia

Dear Shirley,

Our thoughts have been with you and your family since learning of Herb's passing. What a deep tragedy for you and his other loved ones and, to a lesser degree, but never-the-less, a real tragedy, for those of his colleagues who had great respect and much affection, even love, for Herb!

During 1960, I was an over-age postdoctorate in Joe Gall's lab at Minnesota (having spent time in service and then tobacco-farming for six years before going back for a Ph.D.) when I stood in awe of the famous J. Herbert Taylor, passing through to visit Joe after his classical ^{3}H thymidine-labeled chromosome experiments. Soon after, I crossed his path as we both walked to a lecture at a Canadian meeting and hurriedly laid out a far-fetched electron-microscope experiment to also demonstrate single DNA-strandedness of chromosomes. He casually tossed off to me that Meselson and Stahl had pretty well already shown this with bacteria, thus opening the eyes of an eager postdoctorate to the exciting breadth of molecular biology.

Imagine my excitement some years later when Herb invited us to come to Tallahassee to present a seminar, and then later requested slides and photographs of my work to use in his lectures and a publication. There was another trip to Tallahassee, and then much excitement in our research group at the University of Virginia over the accolade of an invitation by Herb to contribute a chapter to one of his *Molecular Genetics* volumes! When my former students arranged an "Oscarfest" for me in 1995, I considered the letter Herb sent me to be a real honor! (A copy is enclosed.)

Shirley, we will not be with you in Tallahassee on February 6, but our thoughts will be with you as you and others remember Herb's beautiful life!

Cordially and with much sympathy,
Mary Rose and Oscar Miller

Herb's message written to Oscar Miller on the occasion of his retirement, for the "Oscarfest" at University of Virginia

June 1995
Oscar L. Miller, Jr.
Department of Biology
Gilmer Hall, University of Virginia
Charlottesville, Virginia

Dear Oscar:

I am sorry I cannot be there to help you celebrate retirement from my Alma Mater. However, it is so changed that I would recognize only the unchangeables, such as the Ranges, Cabell Hall and Jefferson's unique brick walls. Have you ever been to the Blandy Experimental Farm near Boyce and Winchester where I studied? It is much the same as I remember it, except for the program, which is now ecology. The program outlined for your celebration is very impressive. Little did I know that the former tobacco farmer that I first met as Joe Gall's postdoctorate would end with such an impressive and distinguished career. Two former botanists, one an Oklahoma dust bowl farmer and a Carolina tobacco farmer, have come a long way as geneticists after a late start in science. I am a little older and therefore I had the privilege of voting for you and your postdoctorate, Steve McKnight, to be members of the National Academy of Sciences, well deserved honors. All those meetings we have shared with other friends like Barbara Hamkalo, Joe Gall, Aimee Bakken, Don Brown, Steve McKnight, and Mich Kallan among others are lasting pleasures to remember. Shirley and I wish you and Mary Rose many years of happy and productive retirement. My only advice is to do only those things you love to do whatever they may be.

I will be in Australia sailing in the rivers and bays of the Kimberley Coast in the far northwest where the huge saltwater crocodiles prowl surreptitiously in the shallow waters, the great white sharks show their dorsal fins in the deep salt water, but the barramundi strike lures cast into the still waters of the bays and rivers. This is part of a trip to visit our daughter, Lynne, with Mycol and two of our grandchildren,

Brendan (8) and Freya (2), who live in Cairns where we can also visit again the Great Barrier Reef. In July we will take a trip to Siberia by flying from San Francisco to Khabarovsk located on the Trans-Siberia Railroad. We spend 2-3 days on that train to Irkutsk, the origin of a cruise on Lake Baikal to study the unique animals and plant of the world's deepest fresh water lake and see the edges of the Great Siberian Taiga with a Russian limnologist and a botanist as guides.

Congratulations on a great career in which I am sure you found great joy. Your retirement party should be impressive as well as fun. I regret that I will miss it.

Sincerely yours,
Herb

◇

Joseph G. Gall
Carnegie Institution
Baltimore, Maryland

Dear Joan:
Thank you for your e-mail. I am sorry to hear of Herbert Taylor's death. Although I had not seen him in some years, I was an early admirer of his and knew him quite well from various scientific meetings. I became acquainted with his early work on RNA and DNA synthesis in plant material even before his famous experiments on the semi-conservative replication of chromosomes. Those experiments had a profound influence on me—and on everyone else interested in chromosomes—because they provided the first solid evidence that the chromatid consisted of a single DNA molecule. If I remember correctly, his work on chromosomes actually predated the Meselson-Stahl experiment that demonstrated the semi-conservative replication of the DNA molecule itself. A curious sidelight on those first experiments, if you look at his original paper, is that Herbert seemed to favor a rather complex model of chromosome structure that involved multiple strands, rather than the now standard "unineme" interpretation. It is hard for people now to realize just how strong was the prevailing view at the time that chromosomes were multi-stranded. My own investigations of lampbrush chromosomes had led me to the some-

what heretical view that they might be single DNA molecules and I was delighted by Herbert's findings. I was encouraged to try to prove this for lampbrush chromosomes and shortly thereafter did so by showing that the kinetics of breakage of lampbrush chromosome loops by DNase followed second-order kinetics. Another extraordinarily important technical aspect of Herbert's paper was the introduction of tritiated thymidine as a precursor for DNA in autoradiographs. I think tritium had been used only once or twice in biological systems—and again if my memory is correct—Herbert and his collaborators at Brookhaven had tritiated thymidine synthesized, so that they could do their experiment. This provided the much needed resolution for the autoradiographs. Walter Plaut and Dan Mazia had tried somewhat the same experiment previously with carbon-14, but carbon just didn't have adequate resolution. Within a very short time, tritiated thymidine became available commercially, and I remember eagerly buying some and repeating Herbert's experiments to demonstrate to my class in cytology. I also used it to demonstrate that the "reorganization" band that traverses the nucleus of *Euplotes* and other hypotrich ciliates is a wave of replicating DNA. Shortly after Herbert's experiments were published, I invited him to the University of Minnesota to give a seminar—I was a junior faculty member at that time in my first academic position. I remember his lecture very well. He talked about various aspects of chromosome structure and metabolism and then at the very end—almost as an afterthought—described the experiments with tritiated thymidine. I never completely understood why he did this, except perhaps his great modesty prevented him from overly emphasizing his own work. It could not possibly have been from his not recognizing how important the work was! I think he simply felt that he had been asked to talk about chromosome structure in general and that there were many other things that needed to be mentioned.

As you may or may not know, I was "cover editor" for *Molecular Biology of the Cell* for five years, my job being to produce once a month a suitable image and analysis of some historical finding in cell biology. I had decided to do a cover on Herbert's tritium thymidine experiment, but had just kept postponing it. I am sorry now that I did not write him last summer as I had intended asking for a suitable original photograph to use. Could you possibly find one and send it to me? What I would really like is a color photomicrograph showing an autoradiograph of labeled and unlabeled chromatids in a Feulgen preparation. Any help would be much appreciated.

Sincerely,
Joe Gall
(Past president, American Society of Cell Biologists)

Gordon Lark
Department of Biology
University of Utah
Salt Lake City, Utah

A young colleague whose research took a turn after a talk with Herb

Fourteen years ago, in the winter of 1985, I suffered a loss of memory occasioned by the side affects from an overdose of medication, improperly prescribed. Most of what was lost eventually returned, but some aspects of my past remain hidden. Until recently, one such memory was my first interaction with Herb Taylor. It took a conversation with Joan Hare to sweep away the amnesial cobwebs. My acquaintance with Herb began around 1960 when I was an assistant professor at St. Louis University. Since 1954 I had been interested in understanding why cells divided when they did—what controlled cell division. Eventually that was transformed into the question of what controlled, or regulated, DNA synthesis. My ideas were very diffuse, and the only focus was my conviction that this was a central problem of biology. Believe it or not, this was a conviction not shared by many of my colleagues.

I used to say that Herb discovered me, but the truth of the matter was that he forced me to discover myself. He did so by inviting me to write a chapter for the first volume of his book on molecular genetics and encouraging me to pour all of my ideas into that review. I had never done anything like this, and it forced me to collect and focus information and ideas which now seem hopelessly obsolete but which formed the pedestal on which much of our future research was built.

Part of Herb's gift was inborn—a naturally thoughtful, laconic, nature coupled with a very wry sense of humor. Our association was one of several in my life to which I owe much. All of these were role models, often only understood later, but greatly influential. He stands in a very fine company that includes such others as my own father; my graduate mentor, Mark Adams; and associates of my early professional life such as Leo Szilard, E. F. Racker, Max Delbruck, Al Hershey and Taj (Niels) Jerne. Each was characterized by some specific attribute, whose imprint later returned to help and influence science, and me, greatly enhancing my enjoyment of life in general in particular. Herb's quest for truth was slow and thoughtful and always honest. No idea was discarded quickly, none accepted easily.

Over the years we met infrequently, but with that informal ease that allows one to take up a conversation begun years earlier as if it had been yesterday. I remember a visit of his to Manhattan, Kansas. We were sitting on a balcony overlooking our yard and the distant valley of the Blue River. Suddenly a young, newly antlered deer trotted across the yard. As I began to get up, Herb took hold of my arm and quietly said "wait, don't move." Sure enough, about a minute after the deer had left, a coyote came trotting along, following its trail. It is illustrative of Herb's desire to see simple hypotheses tested; it became the first of many examples in which I got to know Herb as a lover of nature and the outdoors.

One of my clearest memories (and one that was never lost) was of a trip to the Rockies that Herb and I made a couple of years later. It was before a cell biology meeting held in Denver and must have been in late October or the beginning of November. We drove to Aspen and spent the night, a very cold one, camped at Maroon Lake. Because of the cold we spent most of the night sitting next to a banked fire which he had built and which burned through the entire night. Topics ranged from the mechanics of camping to the mountains that surrounded us and, of course, to science.

Much later meetings were to include trips to Tallahassee, where I met Shirley for the first time, and subsequent meetings with them in parks in southern Utah. Eventually each of us wants to leave some testimony, some imprint on the world we knew. Herb certainly left his on the people with whom he interacted and through them on their students, friends and families.

He will be greatly missed. In grieving for Herb we grieve for a past, not lost, but that will not come again.

(read by Ralph Cook at the Memorial Gathering)

William K. Baker
Emeritus Professor of Biology
University of Chicago
Santa Fe, New Mexico

Dear Joan,

In 1948 I took up my first academic position in the Department of Zoology at the University of Tennessee, where Herb was an Associate Professor in the Department of Botany. In those days genetics and cytology were intimately intertwined, and it was a real stimulus to have Herb Taylor, Gordon Carlson, and Mary Esther Gaulden as colleagues. In those early days we geneticists were inclined to view genes as abstract localized regions on chromosomes that were responsible for inherited characteristics.

But the behavior of abstract entities was not sufficiently satisfying for Herb: he wanted to know the molecular basis of this behavior. His demonstration of the semiconservative nature of DNA replication in chromosomes was one of the major advances of cytogenetics in this century. The mention of this remarkable research does not pay sufficient tribute to the breadth of Herb's biological interest. He also contributed scientifically to the fields of plant evolution and systematics.

And all of us are grateful to him and Shirley for their endeavors in preserving the natural resources of our country. What a pair they made! We last saw them together in Santa Fe in 1996. They arrived in the next-to-the-last van that Herb had refurnished for traveling and camping. That was a van truly dedicated to studying nature in comfort. Stew and I are much richer for having known them for such a long time.

Bill Baker
(Faculty colleague at University of Tennessee, early 1950's)

Taylor with Mary Esther Gaulden, fellow Blandy farmer, and her husband, physicist John Jagger, at Dallas Botanical Gardens, 1994

Mary Esther Gaulden, Ph.D.
Adjunct Professor of Radiology
University of Texas Southwestern Medical Center
Dallas, Texas

Dear Joan,

In September 1942, Dr. Orland E. White, Professor and Director of the Blandy Experimental Farm Plant Genetics program at the University of Virginia, introduced me (a "new student") to "Taylor," a "final year graduate student who could answer any questions about genetics"! Having just arrived in Charlottesville, a wide-eyed not-dry-behind-the-ears college graduate, I had several unvoiced questions. Was "Taylor" really his first name? He was a young man, so how few years did it take to become a "final year student" who could answer all questions about genetics? Was he a genius?

As soon as we got out of Dr. White's office I sprayed him with those and other questions and got quick, straight answers. No, "Taylor" was his last name: Dr. White called all of his students by their last names, probably because it made male and female students equal! He was in his final year because the draft board was breathing down his neck, and he wanted to finish his dissertation before going off to World War II. He denied being a genius. And, in answer to one genetics question, he gave me his succinct interpretation of the two prominent models (with diagrams on the blackboard) of the relation of genetic recombination to meiotic chromosome chiasmata (an issue being heatedly debated at that time). The one he favored, because it was "the most logical," was eventually shown to be correct, in part because of some of his later research on chromosomes. I shall ever be grateful to him for a splendid introduction to graduate studies!

Impressions gained from that first meeting endeared him to me; they proved to be characteristic of his academic career. He was a kind, generous man, without pretense and devoid of outward animosity toward anyone. It was pure delight to discuss a hypothesis with him, because he became an enthusiastic participant, always open to a different approach or a serious factual challenge to an idea. I remember vividly that when he was asked a question requiring some thought (one on one or in front of students or peers) he would cock his head, focus his eyes on a spot on the ceiling, pause a bit, and then give a considered reply, sometimes as simple and straightforward as "I don't know the answer to that."

Taylor was not only smart and competent in the laboratory, classroom, garden, and carpentry shop but also in his personal life, as illustrated by his marriage to Shirley Hoover, a graduate student in zoology. Their long, happy life together and their family serve as exemplary models for young people. They proved that it is possible to have two successful careers at the same time: science and family. Theirs is truly a great love story!

James Herbert Taylor will be sorely missed not only by his family but also by the many students, colleagues, and friends on whom he left an indelible imprint.

Mary Esther Gaulden
(Fellow graduate student, Blandy Farm and University of Virginia)

Russell Stevens
National Research Council
Washington, D.C.

Dear Shirley and Joan,

For me, and I rather expect for many, the tendency is to reflect in what is a markedly anecdotal fashion. There are, of course, the simple facts that Herb and I, both brand new to the University of Tennessee botanical family, shared an office for several years in total harmony. That we regularly set off to the gymnasium for the daily volleyball game, that he and his good wife Shirley actively promoted my pursuit of one of the graduate students, that he served as best man at my subsequent wedding of that quarry, and that the Taylors shared their home in Maloney Heights with the newly weds whilst their own home was being built.

All this was wonderful, and fondly remembered, but when I reflect on Herb Taylor I confess I think of his remarkable capacity to laugh at himself as heartily as those to whom he described the event in question. Such as—the time he fell off his horse on the way to school, was abandoned by said animal and had to hoof it on foot the rest of the way—as I recall it, the horse was at the school when Herb arrived much later. Or when, greatly amused at a cartoon he found in the newspaper he was reading, he reared back in the bed, smacked his head on the frame thereof and opened an impressive gash in his scalp. Or when, vacationing in the west, he discovered that a VW bus, with throttle to the floor, remains virtually motionless against the impressive headwinds of the Great Plains.

Russell Stevens
(Faculty colleague at the University of Tennessee, early 1950's)

Former graduate students gathered to celebrate Taylor's retirement, 1990.

Memories of Herb Taylor as Mentor – from Scientists Who Trained in His Laboratory

When our life comes full circle and we look back on where we've come from, then we finally understand the true meaning and value of being touched by the brilliance and humanity of a man like Herbert Taylor.

Hervey Cunningham Peeples

Sandhya Mitra, Herb Taylor, and Effat Badr at the International Congress of Genetics in New Delhi, India, 1983

Sandhya Mitra
New Delhi, India

I met J. Herbert Taylor at the Genes Conference at Cold Spring Harbor in the summer of 1951. He was pointed out to me as a very bright young man from the South. I was finishing requirements for a master's degree in botany at Columbia University in the city of New York. I had already been accepted by Marcus M. Rhodes of the University of Illinois for pursuing a doctoral program in genetics. But my plans were altered after a discussion with my former undergraduate teacher at Barnard College, Professor Donald D. Ritchie. He informed me that a very promising young botanist was joining the Department of Botany in the fall of 1951. He asked me whether I was really anxious to learn or would prefer merely to bask in the reflected glory of an already renowned person. Professor Rhodes was already steeped in several ongoing research projects and was not likely to give me adequate time for me to profit from his association. On the other hand, as I was his first doctoral student, Taylor would logically drive me very hard in his effort to establish himself at this prestigious university.

The chief reason for my wanting to shift to Rhodes was the absence of a doctoral guide in genetics in my department. Although stalwarts such as Professors T. H. Dobzhansky, Leslie Dunn, Arthur Pollister, and Franz Schrader were present one floor up in the Zoology Department, the rules prevented a candidate in one department from working under a doctoral supervisor from another. It was not a difficult choice for me. I opted to continue at Columbia, after assuring myself that Taylor had agreed to accept me as a doctoral student. So much for the history of my introduction to Herb.

Herb made an instantaneous good impression with the students of the first course that he offered. He soon became conspicuous as one who worked round the clock. His car was the dirtiest on the campus! Herb and his wife Shirley had no time for such mundane chores as keeping cars in store-bright condition! Yes, Dr. Shirley Taylor was very much a part of our laboratory scene in those days. So was their first child, often entertained by us in the glasswares trolley.

The first course taught by Taylor was one on cytogenetics. Its laboratory component was vastly different from the orthodox, classical labs offered by the genetics and cytology teachers of the Zoology Department. The Dunn- and Dobzhansky-run genetics lab was synonymous with all-night affairs with prolifically breeding fruit flies. The Schrader and Pollister lab involved a term-long intimacy with two salivary-gland chromosomes. Both these labs kept us up nights; I can visualize those flies and their chromosomes graphically even after so many years.

Taylor's lab, on the other hand, consisted of single-class exposures to the different techniques available at the time for learning about the nature of chromosomes. This approach provided an overall picture of current methods for cytological studies. I was given a choice of one of three areas of research that Taylor was interested in pursuing. He already had an inkling that chromosomes split not during mitosis but earlier in what became labeled as the S-phase. He wanted me to follow another line of approach to see whether this surmise could be further substantiated. He wanted to determine the types of aberrations that were produced if chromosomes in meiosis are broken up by X-rays. Karl Sax, the father of radiation botany, had performed such an exercise with mitotic chromosomes. Meiosis, with the complications of crossing over, was a greater challenge. This became the goal for my doctoral dissertation. In this connection I must mention one of the most respect-earning qualities of Taylor. As the data accumulated he did not once indicate his desire to make a quick assessment in order to see if they were about to support his earlier assumption. This prevented us from making a subjective deduction. As my studies piled up, I could sense Taylor's temptation. I felt the pressure for my professor to subconsciously discard data that did not tally with our desires. But he never did. Luckily for both of us, the results were gratifying to us.

During the period of this work, I assisted Taylor in several studies as a Research Assistant, courtesy of an A.E.C., Brookhaven, grant. Most of these involved autoradiographic pictures of radioactively labeled plant nuclei. Autoradiographic "spots" were counted over target regions for a measure of the degree of incorporation of the labeled precursor of DNA or of proteins. This required meticulous scoring of every spot in the cube of photographic film overlying the target region. Thanks to my unusual photographic memory, I rarely counted the same spot twice. Once Taylor watched me doing this for a while. Two actions were taken immediately. I was ordered to take a walk around the block after counting over each nucleus or chromosome; failure to do so would damage my eyes. Second, I was sent to the Bausch and Lomb microscope people to design etched grids on glass discs that could be inserted into the eyepiece of a microscope. These became a standard requirement for counting under a microscopic lens and could be used by anyone with minimum error. This ability to think on his feet and that of a genuine concern for the convenience and comfort of the researchers were reflected in respect and affection by those who were lucky enough to be associated with him. Another attribute of Herb made a lasting and useful impression on me, a member of a rather feudal upper-middle-class family from India. In spite of being an intellectual professional, he had no hesitancy in soiling his hands! My father, a renowned economist, was a good carpenter and handyman. My folks viewed this as the pardonable freakishness of a brainy guy. Taylor walked into the lab one day with armloads of planks of wood. These were bought cheaply from an upstate dealer and were to be assembled by us into bookshelves to house our books and other paraphernalia. He showed us how to do it. Soon the sprawling orthodox biology lab benches were converted into cozy little private domains for each one of us. We were taught and encouraged to repair and redesign other instruments and fixtures as the need arose. The upshot was that we became self-reliant and often self-sufficient.

The training in self-reliance was extended to the scientific sphere as well. In the late forties and early fifties speculations were rampant about the architecture of the eucaryotic chromosome. There were reports of varying numbers of strands, as well as of a possible coiled structure. The problem arose when one tried to visualize how such structures split so effortlessly, without becoming entangled. At this time Taylor had one new idea after another. To understand how a chromosome coiled up, he had his wife sew two bands of elastic tape on the two edges of a cotton tape. Naturally this composite curled up. My job was to represent this on paper. I was no trained artist and failed to come up with a structure that looked like a coiled spring. I was sent to the Mechanical Engineering Department in search of some know-how in this matter. I returned with a book on engineering drawing that taught me not only how to draw a coil that looked like a coil, but many other three-dimensional figures as well.

At another time—well, there are many other anecdotes that remain memorable to me. The upshot was that the once somewhat hesitant foreign student-teacher relationship was lost in a mutually respectful and friendly alliance that was lasting and fruitful.

My first introduction to the professional world of biology was also through Taylor. Professor Lindsay S. Olive, our teacher of mycology, and Taylor packed up their doctoral students and drove them to the AIBS convention at Cornell University in Ithaca. The AIBS meeting was more complicated than a three-ring circus. The number of topics, sessions, venues, and other activities was mind-boggling, at least to one exposed to departmental seminars or at most to Cold Spring Harbor Symposia. The paper setting out the main conclusions of my thesis was introduced by the venerable Karl Sax. I was told that I was very confident and not at all nervous. If they only knew what butterflies were struggling within me! After answering a few questions from the audience, I became a mere spectator. Questions and answers flew back and forth between Taylor, Sax, Muller, Caspari, and many other luminaries of the time. It was a unique learning experience for me.

Taylor was also instrumental in directing me to the most exciting areas of research that had opened out in the life sciences. When I was interviewed by Manglesdorf for joining him in his search for the origin of maize (corn), I was quite excited. This would mean traipsing all over Guatemala and nearby regions and seeing places I would possibly never see again. Manglesdorf's only hesitation was whether I would be able to trek around in my Indian dress, the sari. Taylor felt, however, that I might become sidetracked from the main thrust areas of the newly opening stream of biology. I was subsequently offered two positions—one by Dr. Pappanicoulo of the Cornell Medical School on York Avenue, Manhattan, to study cancer of the uterus, and the other by Dr. A. E. Mirsky, at the

Rockefeller Institute for Medical Research, also on York Avenue, to study the basis of protein synthesis. Taylor recommended that I join Mirsky and keep my ears and eyes open, as momentous things were brewing in the various nooks and corners of that institution.

Sometimes, things were beyond my depth. At such times I would make a beeline to Columbia and get straightened out by Taylor. These sessions were invaluable for me. Taylor represented several different classes of mentors and guides. At the academic level, his quiet insistence on nothing short of excellence, his examples to make us think objectively and not be diverted by tempting conclusions, and his non-imposing but persuasive ways of making us self-reliant molded me, at least, into a somewhat similar teacher.

One day in 1953, Taylor wrote the words DNA DOUBLE HELIX on the classroom board. He mentioned that a momentous report had come out in *Nature*. This announcement might well be the harbinger of a sea change in the understanding not only of the process of inheritance but also of the entire field of life science. I remember the inspired look on Herb's face and the sensation of having gooseflesh while listening to him analyze the historic model by Watson and Crick. The entire class was enthralled by the prophetic way in which he speculated about the significance of the double helix model for a plethora of novel investigations in biology. Taylor immediately devised strategies to corroborate the validity of the model. Taylor kept bean stems in a beaker under the light of his desk lamp. Later, squashes were made of bean tissue and the slides stained and autoradiographed in due course. As Taylor's research assistant, I was entrusted with much of the dog work, not quite knowing the goal of this particular series of experiments. I later realized that this was the pilot study for the investigations that were eventually carried out at the Brookhaven Laboratories and that proved the semiconservative nature of replication of the DNA molecule.

It would not be fair to leave out the contributions made by Shirley Taylor during this time. My first Thanksgiving dinner was at their apartment on 120th Street in Manhattan. I had my first taste of mulled cider, baked beans, turkey, pecan pie, and the works during this dinner. I also heard American country music for the first time. Both my husband, Chitta Ranjan, and I have often talked about the gracious hospitality of the Taylors. Shirley (let's hope unknowingly) also came to the rescue of her husband's "lean and hungry helpers." She used to store fresh pecans from the South in one of our spare refrigerators. We would sometimes raid these bottles, carefully rearranging the remaining nuts so that the pilfering would not be noticed easily! This was my first and only complicity in organized crime. But by far the most significant role played by Shirley Taylor in enhancing her husband's career was to play the role of an uncomplaining (as far as we know) spouse when we appropriated her household tools and appliances for laboratory needs. We had a list of such tools required for our work. Taylor did not at the time possess enough grants to satisfy us. So he would look through our (and his) wishful-thinking list and present his wife with some of them on special occasions. Allowing a decent interval, the items were "borrowed" for the laboratory. Needless to say they never went back. This is how we were provided with such invaluable gadgets as hair dryers, autoclaves (large pressure cookers), and so on.

Years later, after bringing up two children (Amitava and Anuradha) and retiring from the Birla Institute of Technology and Science (BITS) at Pilani, Rajasthan, in India, I visited the Taylors at their elegant but unpretentious home in Tallahassee. It was a delightful visit, made doubly memorable by the reversal of roles by Herb and Shirley. Now it was Herb who concocted delicacies in the kitchen, while Shirley kept herself occupied with matters of the environment. It is my great sorrow that now that I am free to travel more or less at will, Herb is no more.

It would not be correct to say that Herb is not with us. I have felt his influence whenever I launched on a new path or direction in my professional sphere. Entrusted with totally modernizing an archaic Biology Department at BITS, Pilani, I successfully undertook to reorganize the genetics, molecular biology, and molecular developmental biology courses, teaching them as well as writing textbooks for them. I also introduced a broad-based laboratory course in genetic engineering as well as an advanced one in the theory of recombinant DNA technology, again writing books suitable for such offerings. Whenever I was stumped, I wrote to Taylor and promptly got a reply to my queries. He had once told me, "Do not be afraid. After all, you are still baking a cake—only the kitchen has altered slightly!"

Yes, the goal in front of us is still the same as in 1950 or even in the earlier centuries. One still wants to know what ticks within the living system and how it manages to do so, so successfully.

Sandhya Mitra
(Ph.D., Columbia, 1955)

Jesse Sisken
University of Kentucky, Chandler School of Medicine
Lexington, Kentucky

I am pleased that I could be here to share, with Shirley and her family and others here, in the sadness of Herb's passing and to help celebrate his life. I am currently a faculty member in the Department of Microbiology and Immunology, College of Medicine, University of Kentucky in Lexington. I was a graduate student of Herb's in the Department of Botany at Columbia University during the period 1954–1957. This was a time when Herb Taylor did some of his most seminal work, and I would like to share with you my perspective of the man and the times and environment in which he worked because, to understand the greatness of a scientific advance, one has to understand the context in which it was made.

Herb had arrived at Columbia in the fall of 1951, as a botanist. The Botany Department was classical botany, and the physical environment was fairly primitive and makeshift, at least by today's standards. There were four students in the lab during this time: Sandhya Mitra (his first student), the late Rachel McMaster, the Joan Kosan, and myself. The labs where we worked were very simple; they had been taxonomy labs in a previous time. The only lab furniture was broad sitting-height slate bench tops and huge filing cabinets that still held the taxonomical plant specimens from the previous lab. In addition to this main laboratory, there was a small room that served as both office and lab for two students. Herb had an office with a small connected lab and darkroom. The only equipment or facilities he had at the time were a microscope, a darkroom, a couple of radioactivity counters, and some staining jars.

It is hard to imagine now just how primitive our setting was at that time. To examine the base composition of DNA, I used to hydrolyze DNA to its bases, separate the products by paper chromatography, then detect the spots by holding the dried chromatogram against a sheet of photographic paper in front of a UV light. The UV light was a regular light bulb sealed in a plastic quart juice container that held a solution that allowed only UV wavelengths to be transmitted. It was an invention of Lenny Ornstein in our department (whose more important invention was acrylamide electrophoresis).

Our container for disposal of ^{32}P was simply a wide mouth, gallon, glass jar that sat on top of one of the plant-taxonomy filing cabinets where Sandhya and I worked. Herb demonstrated that this was not a dangerous situation for us by showing me that a Geiger counter registered no significant number of clicks when held a foot away from the jar. These methods would not be acceptable today, but these were the days before OSHA or university safety committees and before we really appreciated the dangers involved.

The intellectual climate in the department was also not the most modern. There was considerable disbelief, still, of the double-helix model of DNA. I remember presenting a paper at the departmental journal club that had recently been published in *Nature* by men named Watson and Crick. The paper proposed a theoretical genetic code. The paper and the presenter were summarily destroyed by a professor who stated flatly that there was no way a 4-letter code could account for a biological diversity of 23 amino acids. Irwin Chargaff stated to a seminar class that he could not accept the Watson and Crick model of DNA unless someone could demonstrate the existence of an "unscrewase." In this sort of atmosphere, I think Herb found his primary scientific stimulation from people like Arthur Pollister, Franz Schrader, Sally Hughes-Schrader, and Francis Ryan, all of whom were in the neighboring Zoology Department.

But Herb's major tools were his brain and the database contained therein. Those initial and perhaps most important accomplishments were not serendipitous. He had a goal, which was to study the replication of chromosomes. Autoradiography, the development of which he and Shirley further advanced, had been demonstrated to work at the cellular level, but both the specificity and the resolution were too poor to provide the kinds of images he was looking for. He knew that thymidine was a specific label for DNA, and he had decided from his own investigation of the problem that the ideal radioactive label was going to be tritium—so what he needed for his studies was tritiated thymidine, which in fact did not exist. I remember that in my first interview with Herb, he explained this to me and told me that he had tried to get a chemical company to make some for him. Their price however was something like $3,000, a ridiculous amount of money at the time. In desperation, he considered the idea of making some himself. He traveled to Brookhaven National Laboratory on nearby Long Island and talked to a chemist named Walter Hughes, who told him he thought he knew an easy way to do it. As I recall, the very first or second experiment in which Herb supplied the first ever tritiated thymidine to growing *Vicia*

Taylor with his second Ph.D. student, Jesse Sisken, and one of his final Ph.D. students, Karin Sturm, at his retirement party, May 1990.

roots was successful. This amazing success was a testament not to luck but to both his scientific acumen and his persistence.

Not only was this the beginning of the use of all tritiated compounds in biomedical research, but that one compound alone, tritiated thymidine, opened doors to the study of cell proliferation that had not even been imagined then. It is hard to imagine what research in cancer or immunology, for instance, would be like today without Herb's pioneering development more than 40 years ago. The imprint of this modest man's work has been and continues to be enormous.

I remember when he first reported these findings at the AIBS meetings in Storrs, Connecticut. It was a packed, standing-room-only situation in which people squeezed in specifically to hear Herb talk because they had already heard about his discoveries. In a reserved, almost apologetic way of speaking, he indicated at the beginning of his talk that others had convinced him to talk about his more recent findings instead of the talk originally scheduled. Of course this was what they were all there to hear.

Herb was a patient and forgiving man, qualities also shared by Shirley. This was clearly demonstrated to me the night that my wife and I, with our recently born son, came to dinner at their apartment. Of course the baby had to be fed his bottle and burped after every few ounces. But suddenly there was an explosive burp, which anointed not only Mom and Dad but also the

Taylor's new couch, a situation that would have been trying to anyone but somehow was handled with considerable grace by the Taylors. I witnessed Herb's charitable and patient demeanor again when I gave my first talk as a naïve, student in his cell biology class. Electron microscopy was just beginning to be used as a method for visualizing ultrastructure. People had learned to splay out cilia and flagella on grids so that the individual subunits of axonemes could be visualized. From that work, it had become apparent that cilia and flagella were essentially identical all across the world of eucaryotes. I was pretty excited by this work, but after talking excitedly for a while and showing people these wonderful images, I noticed that at least some of the people in the room were beginning to look a little sleepy. So I stopped and asked how much time I had left. Herb's reply was something like, "Well, you have been talking for almost 2 hours." How many professors, including me, would have been patient enough to listen to a student talk so long?

Herb also had a wonderful sense of humor that was all the more fun because it was so understated. Once, when he had been invited to present his findings at the inaugural meetings of the Biophysical Society, I asked upon his return what a biophysicist was. The answer he gave with that little grin of his was, "A biophysicist is a physiologist who can fix his own oscilloscope."

His reputation as a wonderful human being went beyond those immediately surrounding him. I have been corresponding recently with a Danish colleague, Karin Nielausen, who did a postdoctorate with Howard Green in the 1950's when Green was also in New York. This week she wrote me that although she had never met Herb she had always had the impression that Herb Taylor was an honest and decent man, an opinion conveyed to her by Howard Green, who knew him.

"An honest and decent man"—I think we all had that opinion.

It's sad that Herb isn't with us to share in all these stories and reminiscences, but maybe in a way, he really is.

Jesse Sisken
(PhD., 1957)
(Spoken at the memorial gathering)

Joan Dalheim Kosan
Staten Island, New York

Dear Joan,

How fortunate I was to be a graduate student at Columbia University, 1955–1959, in the Botany Department with Dr. J. Herbert Taylor as my mentor.

It was the most exciting time in biology! The structure of DNA was no longer a mystery, and MY mentor had made a significant contribution toward verifying the Watson and Crick model. I had no previous exposure to such high-level research. You can imagine how excited and awed I was to be there at that time.

Yet for all the acclaim accorded to Dr. Taylor during these years, he remained an unassuming man. Even when he was excited, he seemed relatively calm—a true study in understatement.

I can still recall his incredible patience as he guided me in my research. Autoradiography was part of the project, and I remember how he took the time that was necessary until I mastered this new technique.

Dr. Taylor was always available to discuss any problems his advisees were having. When things weren't going right he would review the circumstances and simply advise us to keep trying. He was especially encouraging when Jesse Sisken's experiment exploded and when my lily anthers still failed to complete the entire meiotic cycle in culture solution.

Keep trying was the advice given. So we did, and over the years we made our own, albeit small, contributions to science. I have such fond memories of those years with Dr. Taylor: scientist, gentle man, and gentleman—good person.

Joan Kosan
(Ph.D., Columbia, 1959)

Deepesh De with wife, Mira, and two daughters at home, India, 1988.

Deepesh Narayan De
Professor Emeritus, Applied Botany Section
Agricultural and Food Engineering
Indian Institute of Technology
Kharagpur, India

Dear Mrs. Taylor,

I am deeply grieved to hear about the sudden demise of our venerable Professor J. Herbert Taylor. On behalf of myself and my friends from India, who worked with him and knew him, I send sincere and heart-felt condolences to you and members of your family.

I was perhaps the third student who had the great fortune of having him as the Ph.D. advisor. Right after his sensational work in 1956, once while returning from Brookhaven, he stopped at Cold Spring Harbor to deliver a seminar talk on chromosome replication. There I saw one of the most brilliant cell biologists, and I asked him to accept me as his Ph.D. advisee at Columbia University.

The years 1957–1960, while I worked for my Ph.D., were the wonderful days when we basked in reflected glory during the visits of Crick and Delbruck, Pontecorvo and Luria, Kornberg and Hershey, and all the molecular biologists and future Nobel laureates to the seventh floor of Schermerhorn Hall. Our working days went into nights, semesters to years through the long snowy winter down Amsterdam Avenue to a brief summer of a couple of picnics at Jones Beach, a few games at the hard tennis court next to Pupin Hall, and an occasional dinner at the Taylors'. But all these were centered on the first-ever tritiated amino acids and nucleic acid precursors. I vividly remember how our professor personally put a sensitive radioactivity monitor inside a refrigerator to obtain a forerunner of the scintillation counter. He has been a great example of innovations and creativity.

After I left Columbia in 1961 and settled down at the Indian Institute of Technology, Kharagpur, in 1962, Professor Taylor's publications kept me on the track of the advances in molecular biology. But the most enduring memory of him I have is his ever-smiling face. As Ellie Goldstein, Jeanny Tung, Dorothy Pfeffer, Ben Bouck, and I would be playing card games or chess during the lunch break, he would stand by for a while with a fatherly smile on his face. The last time I met him, along with you, was at the Birmingham (UK) International Genetics Congress in 1993. Even on his globetrotting schedule he remained excited by the latest developments in molecular genetics. It is a pity that my desire to take him to a "Balti" dinner in Birmingham remains unfulfilled.

Even when the candle is out, the wick glows for long, the fragrance lasts forever.

May his soul rest in peace.

Sincerely,
Deepesh De
(Ph.D., Columbia, 1960)

Barbara Boyes
Research Scientist in Genetic Toxicology
Health Canada, Food Directorate
Ottawa, Canada

Dear Joan,

I came to Dr. Taylor's lab at Columbia, fresh from a B.Sc. at McGill and with a naïveté that, in retrospect at least, seems profound. Since the catalogue said one could obtain a Ph.D. in three years, that's how much time I expected to take! So I crammed in all the necessary courses in the minimum possible time.

While I don't like to think there was any connection, my arrival at Columbia coincided with a number of departures, on the part of the Genetics faculty in particular. Dobzhansky went to Rockefeller, Dunn retired, and a year later Ryan died and Dr. Taylor left for FSU. I had intended to go to Tallahassee to complete my degree there but instead went just for the summer of 1964 and heeded the call of matrimony that fall.

My new husband was a student at Toronto, so like the good academic wife I was at that time, I supported us there for two years by working as a technician and doing my own research in all too few spare moments, as well as in Tallahassee during the summer of 1965.

When both my husband and I got scholarships, we lived on those while I continued my Ph.D. research (for Columbia) under Dr. Taylor's direction, in the laboratory of Klaus Rothfels in Toronto. This bizarre scheme proved unworkable, so eventually I transferred to the Ph.D. program at the University of Toronto, under Rothfels' supervision.

My thesis topic was the unit of replication in DNA of the Amphibia, which very conveniently had extensive differences between species in the amount of DNA per cell. We reasoned that if the replication units were longer in species with more DNA, this would be evidence in favor of the chromosomes being one long strand rather that polymeric as suggested by Shelley Wolff. After much labor, I demonstrated, using ^{3}H-thymidine DNA fiber auto radiography, that the unit is the same length in four species with an order of magnitude range of DNA per cell. While this did not settle the uninemy/polynemy question, it did show that the amphibian DNA replication unit was about the same size as that in mammals.

While all this was in progress, I gave birth to Andrew in 1969 and moved to St. Louis, MO, in 1971. Needless to say, my husband had a job at St. Louis University. Therefore I did the analysis of my experiments partly at Washington University (where I had a teaching assistantship one semester and a lectureship the next) and finished the job while teaching full time (20 contact hours per week!) at Maryville College of the Sacred Heart. So I finally got the Ph.D. in 1972, ten years and many adventures after I began! It was published in 1975 in *Chromosoma*, where Dr. Taylor was editor.

At Maryville I was Science Area Coordinator after the first year, in addition to teaching Human Anatomy and Physiology to nursing students. This I did on the basis of a 2-credit night course taken at Columbia and a lot of chutzpah. We also both taught summer courses at York University in Toronto. I somehow found time to give birth to Dorothy in 1974.

In fall 1974, my husband began a permanent full-time job at York, so I applied everywhere within two hours' drive and landed a job at the University of Western Ontario, Zoology Department, London, Ontario. We brought a house there, as his work was more portable than mine. A year later I moved to the Anatomy Department; they offered me a lot more money as well as a tenure-track job. I thus found myself teaching medical students on the basis of more chutzpah than I knew I possessed, as well as frenzied cramming far into the night! In addition, I was beginning a series of experiments on combinations of mutagens, using as endpoint micronucleus frequencies in tissue cultures, to which I had been introduced previously by John Heddle at York. The breakup of my marriage during this time also complicated my life.

During my years at Western, I taught a lot—I got pretty good at anatomy after a while—which didn't leave nearly enough time for research. In 1983 I moved to Ottawa, where I am Research Scientist for the Federal Government in Genetic Toxicology. What a change! It's like full time sabbatical—no teaching (I do miss it), little committee work, and adequate money for research. I continued the micronucleus work, gradually switched to sister chromatid exchange, and am now happily in the midst of various projects using an EPICS 752 flow cytomoter/cell sorter.

Thus almost all my professional life has been involved with one aspect or another of cytogenetics, although, curiously enough, I have never prepared a karyotype! For the flow cytometer work, I have had to learn a certain amount of immunology in addition to how the machine works. It is sobering to realize how much of what I do was not even invented fifteen years ago.

Certain things I admired in Dr. Taylor I have tried hard to develop in myself; particularly the clarity, conciseness, and logical organization of his writing, the sheer fun of working in the lab, and the careful way in which an idea is worked out and vigorously tested. I am very grateful for the excellent start he gave me in the exciting and demanding career of scientific research.

Barbara G. Boyes
(M.S., Columbia University, 1964)

Marvin Rosenberg
Professor of Biology
California State University, Fullerton

Dear Joan,

As you now realize, I will be unable to come to Tallahassee for this Saturday's Memorial. Our daughter has been seriously ill, and we have been totally occupied as parents and grandparents. The semester started this last Monday, and I have so much catch-up.

I did send a letter to Shirley, to which she has already responded, and also made a donation to the University Arboretum, and Herb's name will be permanently placed in the Memorial Volume. I know that both Herb and Shirley would have loved the Arboretum if they visited. It is a small space of beauty and serenity in the midst of an urban sprawl.

The greatest testimony to Herb is that he accepted me into the lab at Columbia. I was a secondary-school teacher who had just switched to a college position at the State University of New York, then at Oyster Bay, now at Stony Brook. Although I had a master's degree in botany from Cornell, it represented a risk. In 1960, the Botany and Zoology Departments were separate entities, and I believe Herb held appointments in both (Shirley can confirm this). At that time, Columbia biology was excellent and had many well-known and acknowledged faculty. The papers on semiconservative replication of chromosomes were newly published and provided cellular verification to the Messelson-Stahl work. I worked at the Cold Spring Harbor Labs, and so many of the investigators suggested him as a possible mentor. Amongst them were Milislav Demerec, Barbara McClintock, Francis Ryan, Al Hershey, and other early molecular workers.

After completing my course work at Columbia, we came to the Institute of Molecular Biophysics in Tallahassee, where I completed the research, and returned to the Stony Brook position. Herb told me about and also recommended me for a variety of positions for which I interviewed. I accepted this job at California State University, Fullerton, in 1968 and have been here since. My graduate students now hold positions at campuses such as Wisconsin, Georgetown, Berkeley, and others—these are the third generation in the Taylor family. I was a research associate at Cal Tech in Pasadena for two years on sabbatical, worked in the Davidson lab, and cloned (with the group) the sea urchin gene for myosin and demonstrated the coordinate regulation with actin. I was chair of the Biology Department for six years, Associate Dean for eight years, and Acting Dean of the School of Natural Science and Mathematics after the unfortunate death of the dean. I accepted early retirement in 1993, which means that I teach only one semester a year and have turned my attention to science education at the elementary level. I have been awarded grants in excess of $1.5 million for this work.

I could not have had this lifetime career I've described without Herb. He was such a gentle, brilliant, dedicated man. I owe him a great deal.

Please share this letter with Shirley. I hope that her memories will sustain her through these difficult times.

Marvin Rosenberg
(Ph.D., Columbia, 1966; FSU 1964–1965)

Taylor reaching into a bog-garden with Bill Haut and Bob Hopkins at their Stillwater Farm, Dewart, PA, 1988

William Haut
Sierra Vista, Arizona

Dear Shirley,

We have just returned from holidays with family to find your sad news awaiting. We will miss him. We do live on in the hearts and minds of those who loved us—and he was loved by a host of people ... a most gentle soul.

I am thankful that our numerous visits with you and Herb in Pennsylvania in our old farm house, and our times together in Florida, gave me ample opportunity to share with Herb my profound admiration and respect for his abilities as a scientist, teacher, and intellectual journeyman. He was my mentor. He was my friend. His willingness to take me on as a graduate student and guide me to a Ph.D. made it possible for me to experience a wonderfully satisfying professional life. For this I am deeply grateful.

The last time that you and Herb visited us in Pennsylvania was late summer, 1997. Herb had just finished your new Ford van conversion, and the Polar Bear was bound for home after completing its first road trip. During that visit I found Herb to be particularly philosophical as he shared endless memories of his childhood in Texas, his youth, and his years at Blandy Farm, where he had known my undergraduate professors at William and Mary, John Baldwin and Bernice Speese. Baldwin and Speese were an important part of the reason that I ended up at Columbia in Herb's lab for my graduate work.

I shared with you both during that visit some comments that I had made at the memorial service in May for a colleague that I lost to cancer. I had spoken about my personal heroes: Heroes defined as "persons of distinguished ability and courage, admired for deeds and noble qualities." Herb was on that list. "My heroes share several commonalities. They were my teachers. They loved life. And they were passionate. Passionate about their art, their science, their spiritual values, their parenting. Passionate about being compassionate and caring. Passionate about teaching."

Herb was on an endless quest. Original discovery was intensely important. Research for him was not a sport that satisfied personal ambition, it was an art form. I remember his eyes bright and sparkling with anticipation as he searched the data for an explanation. And I remember him, glasses in hand, eyes tightly closed, quietly searching his imagination for questions as much as answers. Like an artist with an empty canvas, he kept painting new pictures with searching creativity, talent, and a willingness to risk. How fortunate I was to know this man both as mentor and as friend.

I will miss him, my caring teacher, and my hero, who loved his science with such passion. He was a truly wonderful human being in every measurable aspect, an inspiration to his students, and for me a caring mentor and gentle guide who helped me steer a mindful course whenever the going got tough. He enjoyed a long and full and meaningful life, touching many along his path. How wonderful!

Bill Haut
(Ph.D., Columbia University, 1966; FSU 1964–1966)

Postscript: Do you and the family have any plans to establish a scholarship at FSU or some other memorial/tribute in Herb's memory? Please let us know as plans unfold: We would like to contribute in some way to a lasting tribute to his spirit and mindfulness.

J. Kenneth Shull, Jr.
Professor, Department of Biology
Appalachian State University
Boone, North Carolina

Dear Joan,

I had heard about Herb through Terry Ashley, and I will forward your message to Dwayne. I am really very sorry to hear about his death.

When I was a graduate student (M.S.) at Alabama, I got interested in chromosome structure and was assigned to give a seminar on that subject. Obviously, I ran into Herb's name in the literature, since he had published several models of chromosome structure. When I learned that he had moved from Columbia to F.S.U., I applied for doctoral work there. Although I did my work with Margaret Menzel, it was because of Herb that I became interested in chromosome structure and went to Florida State.

Herb was kind enough to address one of the Boone Chromosome Conferences that I host here at Appalachian State a few years ago. He proved to be as insightful as (and, in some ways, a better speaker than) he was 17 years earlier.

We will all miss him. Please convey my condolences to Shirley and the rest of his family.

Thank you for your message.
Ken Shull
(Ph.D., FSU, with Margaret Menzel, 1973)

Terry Ashley
Associate Research Professor
Department of Genetics
Yale University

Dear Joan,

I was the first FSU student to join Taylor's FSU lab—in the fall of 1964. He had just moved down in the spring. The Molecular Biophysics Building had just been completed and was not fully occupied. The lab was very isolated. Bill Haut, one of Herb's students from Columbia, was there when I arrived. He had brought his technician Jean Tung (Lee) down from Columbia as well. They were followed around Christmas by the late Philip Minor and later by Marvin Rosenberg (both from Columbia). Helen Crouse also moved down from Columbia after Christmas. For a couple of summers, Barbara Boyes (Wilson), yet another Columbia student, joined the lab to work on her thesis. In those days before the safety regulations of today, we ate lunch in the lab. Barbara was a real live wire. When she was there, there were midnight trips to the beach (by us graduate students), and there was even a weekend when Barb, her husband Barrie, and I "went to lunch" one Friday afternoon and returned after lunch on Monday, having made a slight detour to the Keys in the meantime!

Herb was trying to make the transition from cytology to molecular biology. He did not have a lot of the (then) modern equipment, so I spent much of my time counting drops and moving test tubes, because he did not have an automatic drop counter! I also did some (more cytological than molecular) autoradiographic work on grasshopper testis—my first meiotic work.

We used to joke about Taylor's Replication Models. At that time, they were like cars—a new model every year. However, I think we all appreci-

Dwayne Wise and Taylor at the Second Boone Chromosome Conference, Boone, North Carolina, 1989

ated the amount of thought and effort that went into these endeavors. He really tried to integrate everything that was known about chromosome behavior and replication into each current model. He ignited an on-going interest in DNA replication for me.

As you probably are aware, I have worked on mammalian meiosis since the late 70's. One of my favorite quotes from Taylor was his ending to an introduction to a session on meiosis he gave at an International Genetics Congress in Berkeley. The proceedings were published in *Genetics* 78, 1994. He said "I will only remind you that meiosis is a potential battleground where dead hypotheses litter the field or rest uneasily in shallow graves, ready to emerge and haunt any conscientious scientist who tries to consolidate a victory for any particular hypothesis." As I have struggled to understand what is going on in meiosis, I am continually reminded of this admonition.

I switched thesis advisors to Margaret Menzel. Several of her students started a Chromosome Conference in Boone, NC, several years ago. Dr. Taylor and Shirley were strong supporters of this conference, and we all have appreciated their on-going support.

Terry Ashley
(Ph.D., FSU, 1970)

Dwayne Wise
Professor, Department of Biological Sciences
Mississippi State University

Dear Joan:

Here are some of my thoughts about Herb Taylor ...

I remember the first time I saw him. I had signed up for a course called "Advanced Genetics," taught by someone named Taylor. I went to class the first day, in Unit I. A tall, slender, slightly stooped, very unassuming man was standing in the corridor. To my surprise this man walked to the front of the room, opened the book near the middle, and began to talk in his quiet way about something very complicated.

I went on to learn much from the course and to get to know Herb Taylor. He was unfailingly polite, but had the most penetrating mind.

He would attend the weekly genetics seminar, appear to sleep through it, and then rouse to ask the one most penetrating question that none of us had considered. Years later, he always knew my name and made a point of speaking to me at meetings, conferences etc.

He was one of the most admirable people I have known, at all times the consummate gentleman. In many ways, he is the model that I strive to emulate.

Dwayne Wise
(Ph.D. with Margaret Menzel, FSU, 1972)

Madeline Wu
Professor and Head, Department of Biology
Hong Kong University of Science & Technology
Clear Water Bay,
Kowloon, Hong Kong

Dear Joan:

This is a short response during my trip to Hainan to work out a collaboration project about the environmental research in South China Sea.

Professor Herbert Taylor's death was a big shock to me. I still cannot accept and respond to this fact properly. I have had so many trips recently, I have had a hard time sorting out my thoughts. I also did not check my e-mail regularly. I have informed Jungrung Wu, my husband, about this, and he was equally surprised. This is such a big loss, we cannot face it properly yet. It is a big comfort for us to communicate with you all to share about this big loss.

Herb was much more than a mentor of scientific research to us. During our short stay in Florida, he and Shirley had treated us as their family members. Through close interactions with the two of them, I gained some real insight into environmental issues, which had a real impact on my future research direction. When I started to develop my own research career, he was also the most helpful person and communicated my manuscripts to the *Proceedings of the National Academy of Sciences of the USA* and *Chromosoma*.

I have always dreamed of joining one of the field expeditions with him and Shirley. This dream can never be realized now. From your description I know he was so brave in fighting the disease toward the very end. My utmost admiration always goes to this great person.

Sincerely,
Madeline and Jungrung Wu
(postdoctoral research scientists, FSU, 1970)

Hervey Cunningham Peeples
Longboat Key, Florida

Dear Joan,

Thank you for sending me word of Dr. Taylor's passing. I was profoundly saddened at the news. I have a strong sense of nostalgia for those great days (over 30 years ago!) at FSU with Dr. Taylor and Shirley that remains with me as I write this.

I remember Dr. Taylor, first of all, for the awe and respect he inspired in me, the challenge he gave me to succeed, and the delight and pride I felt when I first learned he had chosen me out of many new genetics graduate students to work with him on my master's degree. Later I learned about his integrity, his personal warmth, and his chuckling sense of humor ... I can still see him sitting with Shirley at the lunch table in the middle of the lab, watching me don a huge blue parka before entering the cold room, and laughing out loud at the sight since the parka was almost as big as I was, but only as long as my mini-skirt, so my legs hung out unprotected. He couldn't understand why the cold never bothered me—and neither could I!

But above all else, I remember Dr. Taylor's limitless patience with me. Though I had the privilege to study with him only a couple of years, it was long enough for those seeds of self-discovery (AND self-confidence) to be planted, and they're still growing. ...

I'm taking the GRE next month and have plans to return to graduate school in a program combining my lifelong love of genetics with a recent fascination with anthropology. As a result, I have thought of Dr. Taylor many times over the past several months, so the sad news was made even more immediate. I guess it's times like this, when our life comes full circle and we look back on where we've come from, that we finally understand the true meaning and value of being touched by the brilliance and humanity of a man like Herbert Taylor.

Please accept my sincere condolences on your loss, and my appreciation for keeping me "in the fold."

Hervey Cunningham
(M.S., FSU, 1970)

Dr. Joseph Adegoke, in his garden on sabbatical in Sweden, 2004

Joseph A. Adegoke, Ph.D.
Professor of Genetics, Department of Zoology,
Obafemi Awolowo University,
Ile-Ife, Nigeria.

May 2004

Dear Shirley,
Thanks so much for your stimulating and all embracing mail! I read

it over several times and felt like I was hearing you talk. I wonder where those letters could have been directed because we got none of them. Of course in the 1990's it was usual for the postal workers to steal just about any mail coming from overseas, especially USA.

I would certainly love to write something in memory of Professor Taylor: My first contact with Professor J. H. Taylor was in May 1970, when the USAID's African Graduate Fellowship Program (AFGRAD) awarded me a fellowship to pursue studies leading to a Ph.D. degree in genetics at the Florida State University, Tallahassee, with Professor J. H. Taylor as my supervisor. I was exceedingly excited, since I had strongly indicated my desire to study the structure of eukaryotic chromosomes when filling in the AFGRAD forms.

I arrived at FSU in mid September 1970 and met Dr Taylor for the first time when I brought my registration forms to him. Having seen several of his publications and his two books, *Molecular Genetics I* and *II*, I felt very elated to meet him personally and to be one of his graduate students. My graduate-studies program spanned exactly four years. During this period I never saw him get angry or complain about this or that. He regarded everyone in the lab as members of one family, advising students regularly on academic matters and quite often also on personal matters. He was never overbearing. One was able to learn a lot of experimental procedures, which have proved very useful in my case, as I could adapt some of the primordial facilities at Ife to obtain publishable data. I regard myself as having been fortunate to have the opportunity to work with such a great and gentle, but very alert, mind, whose dedication to work is superlative. His works have stood the test of time and will remain valid for posterity. He has been a role model for me in my research and interaction with my graduate students.

My family and I enjoyed our brief one-day stay in the home of the Taylors in Tallahassee in the fall of 1982, when we visited. The pictures we took with them then are treasured by every one of us and are firmly glued into the family album. Finally, we thank God for the life of Dr. (Mrs.) Shirley Taylor, whose energetic doggedness has made this important publication possible.

Joseph Adegoke
(Ph.D., FSU, 1974)

John Hozier, Herb and Shirley Taylor, Lynne Wall, Jed Dillard and Joan Hare at Angelo's restaurant, Panacea, Florida, May 1986

◇

John Hozier
President, Applied Genetics Laboratory
Melbourne, Florida

Dear Shirley,

I'm sure you know how sad I am (and so many other people are) about Herb's passing. Joan read to me (over the phone) some of the things other people have said about him, mostly having to do with his impact on science, or some particular moment that sticks in the mind.

All that is true for me too—especially his influence on my professional life. And the stick-in-the-mud happenings as well: like getting a phone call early on a Saturday morning—to go fishing with Herb in the Gulf of Mexico, in a canoe, seemed strange at the time. And seems now, since I can't remember any of the details, except that we didn't capsize in the swells, and that it was fun.

But Herb's influence on me covered more than the science: he was the first grown-up I knew at all well who seemed to get through life and enjoy it based on his own inner resources. I remember going to a Cell Biology meeting in Miami with Herb a few weeks after my Ph.D. comp exams—I'd passed; only two of our class of seven did. We stayed in a motel near the meeting site. Herb got through the whole hectic day and evening of events—and still seemed calm and at ease.

I've never forgotten the very normal aspect of Herb's life, and remembering has helped me greatly—reminding me that it's possible to be excited about what is going on in and around you, and at the same time be capable of serenity.

Anyway that's the way Herb's life seemed to me—exciting and serene. Hope that is close to the truth. He certainly helped change my life for the better.

Sincerely,
John Hozier
(Ph.D., FSU, 1975)

Roger Clark
University of Denver
Denver, Colorado

Dear Joan,

I remember many things from the brief two years (1971–1973) I was in Herb's lab at the Institute of Molecular Biophysics. Originally I went there to do an experiment regarding meiotic prophase DNA synthesis. Not long after arriving, I realized that probably this experiment could not be done there. But there was not as much discouragement in this realization as one might think. There couldn't be a better place to learn about chromosomes, and this fellow had done one of these experiments with chromosomes that forever set the direction of all that follows. Such experiments with chromosomes (or anything else) are rare. I thought I would learn a lot from him, and this all proved true. And I recall the lab's chase after replication "subunits" and "30 micron pieces," that surfaced for me years later when my lab pulled such things from mitotic chromosomes.

For my family some of our fondest memories are from the time we spent in Tallahassee. Our first child was born within days of our arrival in December 1970. Even she remembers living there—the beach and the sea—and she was not quite two years old when we left. We haven't been back—the demands of life have interceded for what now has become 26 years. But my memories over that time have not really dimmed much.

Mostly I remember what it was like to learn from this man who is now gone. It was and is a privilege.

Roger Clark
(postdoctoral research scientist, 1971–1973)

Len Erickson
Robert Wallace Miller Professor of Oncology
Indiana University Cancer Center
Indianapolis, Indiana

Dear Joan,

I have several fond memories of Herb, some scientific, and some personal. The first was shortly after arriving in Tallahassee. We were all sitting around the table having lunch family style. I asked Dr. Taylor if he was going to the big football game on Saturday. He got an odd look on his face and that twinkle in his eye and told me "I don't care what that silly football team does!" I soon learned that although quite a sportsman, he did not care for organized sports.

The second memory I have is sitting in the lab with Dr. Taylor at 2:00 a.m. waiting for our synchronized CHO cells to enter S-phase so we could do 15-second pulse-labeling experiments to isolate the newly initiated Okazaki fragments. While sitting there he told me how difficult it was for him to go off to college and leave his horse at home.

Another great memory was a canoe trip that I planned and took with Herb, Shirley, her father Dr. Hoover, Bill Mego, and Nancy Straubing. We decided to canoe down a river in a swamp known as Tate's Hell. On the map it looked like an easy six-mile trip. We dropped cars at the south end and started down the river from the north. What we didn't know was that about every 100 yards the river meandered east and west about a 1/4 mile. The day got hot; we saw water moccasins and alligators everywhere, and we had to portage over many downed trees always wondering if there was a snake or gator on the other side. Dr. Hoover was getting sick with heat exhaustion, and Herb and I got quite a bit ahead of the others. We stopped and he said he was going to swim in the tea-black water. I said what about the snakes and alligators? He replied "the heck with them." We went swimming and waited for the others. We managed to struggle out of the river just as it was getting dark, about 12 hours after we had started! Needless to say, I didn't plan any more canoe trips for a while.

Another fond memory occurred when Pam and I returned from our wedding/Christmas trip in January of 1974. Herb called me into his office and passed a letter to me from Kurt Kohn of the National Cancer Institute. I read the letter and learned that Kurt was looking for a postdoctorate to come to NCI and study the effects of antitumor drugs on DNA replication. Puzzled, I read the letter and asked why he was showing it to me (I didn't think I would finish my Ph.D. for a year or two). Herb said, I've written to him and recommended you for the position. I was speechless. Six months later I started a nine-year stint at NIH, where I eventually was tenured as a Senior Investigator.

The following year Herb called me at NCI and invited me to the public ceremony for his induction into the National Academy of Sciences. What an honor to be remembered and invited to this prestigious occasion.

My most recent fond memory of Herb is when many of us came back to Tallahassee, as a surprise, for Herb's retirement ceremony. We were all waiting on the porch of Joan Hare's home when Herb and Shirley drove up. He didn't have his glasses on but as he came up the stairs he looked at me with that same twinkle in his eyes and said, "I know this guy."

Len Erickson
(Ph.D., FSU, 1976)

James M. Pollock, Jr., Ph.D.
Florida Department of Law Enforcement
Jacksonville, Florida

Dear Joan,

Although I had met Herb at the University of Miami, when he came to lecture at graduate seminars, the first time I really got to talk to him was when he interviewed me for a postdoctoral position in his lab. My very pregnant (seven months) wife and I traveled to Tallahassee while, unbeknownst to us, a hurricane was also approaching from the Gulf. The hurricane—never more than 50 mph in Tallahassee—was there and gone in a few hours. I stayed for five years.

During that initial meeting, the interview, Herb made me feel that I had a tremendous amount to offer, that I had more potential than I had thought I had. I felt comfortable in the area of DNA research, in which I had received my doctorate, of course, but young and unsure of myself in the field in which Herb and his lab were engaged. Herb's positive attitude gave me confidence that I would not only prove competent in this new field, but that I could make a major contribution.

Doing research in Herb's lab was an adventure not only in research science, but also in scientific attitude and the curiosity that drives it. Herb was always open to new ideas and avenues of research. It made no difference if the idea came from a graduate student or a peer professor, Herb always gave the idea his genuine consideration. For instance, in searching for a model of an early replicating DNA, we looked at a range of systems, from fish gametes to pine pollen, finally settling on a sea urchin embryonic system. Even though the fish sperm system was quickly discarded, I hear that the cast net used to catch mullet for research was "filed" in a corner of the lab for many years afterwards.

Most scientists of Herb's caliber and stature seem to be so focused on their research that they have no life or interests outside the lab. Herb was never so limited, and certainly never suggested any of his students or postdoctorates aspire to such limitations. Herb had broad interests, especially in the natural world, and he generously shared his interests and expertise. Whether it was inviting us along on a canoe trip down the local Wacissa River, or suggestions for an off-the-beaten-path camping and hiking area in North Carolina, Herb was a source of encouragement. We first took our son pup tent camping to the Joyce Kilmer Memorial Forest when he was a mere ten months old. Over the years, we have returned to that area many times. When my daughter graduated from high school, almost twenty years after we left Tallahassee, one of the graduation gifts she requested was a backpacking trip. (The trip was so important to her brother that he arranged to take military leave in order to go along.)

And then there was the fishing … One of my clearest memories of

Herb was a time when we set our canoe on a course to the far side of Lake Jackson in search of a trophy bass. Not much was happening, so we stayed until after dark. When we finally decided to head home in the moonlight, Herb confidently gave directions to get back to the car. It was not until we arrived at the northernmost end of the lake that we realized we were somewhat off course. Research is like that, too, I think. The true course is never absolutely known, but one never gets anywhere without choosing a direction and discovering where it goes.

Jim Pollock
(postdoctoral research scientist, FSU, 1974–1977)

◇

Tom Laughlin
Grapevine, Texas

Dear Shirley,

Dr. Taylor has been a great inspiration in my life. To me he was a scholar, mentor, and friend.

As a major professor he was mentor and scholar. He directed my lab work and dissertation writing, with a brilliant scientific approach and a keen technical knowledge. I admired his excellent verbal communication skills. As a mentor and friend, his marriage and family relationships were great examples to us all.

Tom and Darlene Laughlin, 1978.

But it was really the example he set that led the way in my life that followed graduate school. That example went far beyond the lab work and dissertation writing. He showed me a positive exciting approach to life. He provided me with a vision of how things can be done when you combine knowledge, drive, patience, and compassion.

He will be greatly missed. His inspiration lives on.

Tom Laughlin
(Ph.D., FSU, 1979)

Barbara Boggs
Department of Cell Biology
Baylor College of Medicine
Houston, Texas

Dear Mrs. Taylor,

On Saturday, January 2, 1999, I read in the paper of Dr. Taylor's passing and was deeply saddened. I was not aware of his decline in health and am sorry I did not get to see him again. Both of you have been such an influence on my life, and I would like to thank you for the many enjoyable memories that I have.

I will always treasure the afternoon that Craig and I spent with Dr Taylor discussing some of our research on replication timing in the human X chromosome. During the course of that research there were many times when we would say, "Dr. Taylor would love this," and I was so excited to have a meeting with him to show him the results. I had come full circle in my scientific journey, and I had been so thrilled to bring together my first research experience with him with the molecular techniques I learned later. As we left that afternoon, Dr. Taylor shook my hand and said "thank you." In that simple gesture he gave me much more than I ever could have given him that day. I treasure that moment more than any other in my scientific career. I will never get another handshake, but I will strive for the research excellence that will allow me to say "Dr. Taylor would love this."

The last time I saw Dr. Taylor, Craig and I brought our girls along to visit. You all were planning a trip to Africa, and Dr. Taylor was so excited about it that he brought a map and book out to show the girls where you were going and what kind of animals you were going to see. They of course were thrilled to see animals they had never heard of before. Upon learning of Dr. Taylor's passing, our oldest said, "I remember him and going to his house." I hope in years to come when she reads Dr. Taylor's name in science books she can remember that afternoon and say, "My mom had the honor of being a student of his."

The true magic of being in his lab was not just that he had a high standard of excellence but that he also cared about people, for who and what they were outside the lab. To have stumbled into Dr. Taylor's lab as an undergraduate was one of the luckiest things that has ever happened to me. Dr Taylor will always be my role model in science and in life. I regret not being able to be at the memorial service on February 6. Previous commitments prevent my attending, but Dr. Taylor, you and your family are in my thoughts and prayers.

Barbie Boggs
(undergraduate, 1979–1981)

Lynne Wall
Walton on Thames Surrey
Sussex, England

Dear Joan,

I first met Herb and Shirley Taylor in the late summer of 1982, in the UK. I was in the final year of my first postdoctoral fellowship at the University of Sussex, and I had heard from my own professor that Herb had some funds for a UK postdoctorate to join his lab in Tallahassee. Both labs were working on the factors controlling DNA replication, and it seemed to be the natural next step.

Herb and Shirley came to see me in the lab at Sussex. It was not so much an interview as a wide-ranging conversation on all sorts of topics—scientific, natural history, cultural, personal—with lots of smiles and laughter thrown in. There seemed to be no doubt that I would go to Florida State, and by February 1983 I was there.

Herb Taylor gave me a unique opportunity in going to FSU and to the US. It turned out to be a period of significant personal development for me in which the lab played a very important part. Herb struck that balance between being highly supportive—on all fronts—but allowing at the same time independence and a sense of responsibility to predominate in the lab and in all matters to do with research. His curiosity was infectious, and his incisive analytical approach combined with gentle good humor rubbed off on us all. Being in Taylor's lab meant doing all things with good grace and with a zest for life—

Lynne Wall and her husband, 2002

whether that was picking wild berries, listening to music, visiting the wilder parts of Florida, or speaking up at lab meetings.

In the end I stayed in Tallahassee for nearly two and a half years. I came back to the UK worldlier, more confident and more mature. And a better scientist. (These days I work at the interface between scientific research and policy in government.) I owe this to Herb Taylor for giving me the chance.

Lynne Wall
(postdoctoral research scientist, 1983–1985)

Marcia Applegate Harris
Primary care physician
Stuart, Virginia

My first memory of Dr. Taylor was the day I walked into the Institute of Molecular Biophysics with my book of all the major professors in the Molecular Biophysics Program and their research interests. I had pretty much ruled out retinal physics, quantum physics, and statistical thermodynamics and was looking, well, for something biological. I saw Herb's profile in the book and thought DNA and molecular biology sounded about right, so I wandered up to his door, knocked, and walked in. I said I was a new MOB student, and could I be his student. I'll never forget the huge laugh he answered with, and then he said, well he guessed so. At the time I had no idea at all who J. Herbert Taylor was, or I would never have just marched into his office and asked for a job!

I remember he was the perfect teacher for me. He fed me ideas and let me grope my own way to a project, never critical of my constant meandering off the main track.

I remember how he himself was curious about everything, sometimes to the point of embarrassment. Once he decided he wanted to know what exactly was in warm milk that made people relax and get sleepy. He set up a huge protein fractionation apparatus in the lab and tried drinking the fractions—and to get some of us to drink them too! Everybody was glad when he got interested in something else!

I remember how proud we were when he was voted into the National Academy of Sciences, and how he told us his colleagues at Columbia had warned him not to leave and go to FSU because if he wasted his research life at that obscure place he'd never get in the academy. He'd just said that New York was no place to raise kids and left anyway. That was just one example of his devotion to his family.

He taught us that research is fun, and he had a sometimes wicked sense of humor. I'm embarrassed to say that one of my most vivid memories is his trip around the lab one day with a Geiger counter, showing each of us the loud evidence of the radioactive implants he had chosen to treat his prostate cancer.

He was always a Virginia gentleman and behaved with grace and tolerance even when most exasperated. One of the first things I did as a graduate student was to blow up a high-speed centrifuge—about $2,000 (1976 dollars) in damage. I thought my days in his lab would be few indeed. Instead he just sauntered over, peered at the black dust that was all that remained of his rotor and said, well, he saw I'd had a little accident. That was all I ever heard about the matter. I remember how, in a rare moment of pique, he said something about that being a hell of a way to run a company and how astonished we all were at his brief moment of common speech.

Marcia Applegate Harris
(Ph.D., FSU, 1984; M.D., U. South Alabama, 1992)

Stephen V. Angeloni
Georgetown University Medical Center
Washington, D.C.

Dear Shirley,

Herb was one of the two people at FSU who helped spur my interest in molecular biology and genetics. It is a bit painful not to be able to make it. My sincerest regrets and best wishes to you, Shirley. You and Herb gave me the opportunity to develop my scientific approach, which has allowed me to make significant progress in most of the research. I can't thank you both enough.

As an undergraduate in the biology program at FSU, I wanted to get a taste of working in a research lab and volunteered to provide my services for whatever was needed in the Taylor lab. As is customary for such a young apprentice, I was given the very important task, which no one else in the lab wanted, of emptying and washing scintillation vials. For an undergraduate who wanted to study molecular genetics and see how it was done at the time, I considered this a lucky opportunity. When my "voluntary" services were eventually paid for, it was an even greater experience. It also made me wonder, could you really make a living doing science? After finishing my undergraduate studies, I left FSU for a while then eventually returned to study for my master's degree. During this time I was again working in the Taylor lab to put myself through the graduate program. It was during this time that Herb, Shirley, and the members of the Taylor lab instilled in me the ideals of being meticulous and uncompromising when it comes to doing quality research. I don't know if I can say I've returned the favor aside from providing hundreds of sparkling clean vials and other assorted glassware.

I may have contributed a bit more by helping bring the lab into the computer age. During this pre-internet, pre-Windows period, there was an

Apple II computer sitting on the secretary's desk collecting dust. Out of the necessity for writing papers, reports, and organizing lab inventories and methods, I was able to talk Herb into buying additional hardware and software to get it running for word processing. His secretary at the time, Mrs. Hurst, didn't want to go near the machine. She was content to "cut and paste" the old fashion way, with metal scissors and sticky tape or paste.

Computers and the molecular techniques we use today have come a long way since my days in the Taylor lab, but what doesn't change is the need to plan, prepare, and perform experiments with the exacting care for detail that I learned from Herb and the members of his lab. This approach to science has remained with me to this day and has been instrumental in allowing me to be successful in studying molecular and genetic mechanisms in the areas of prokaryotic, eucaryotic, and cancer biology.

I can't thank Herb enough for giving me my first exposures to molecular genetics and shaping my approach to science. This experience and the ideal of doing the best job possible will stay with me no matter what the future brings.

Thank you Herb for everything. I will remember you always.

Steve Angeloni, Ph.D.
(undergraduate and graduate lab assistant, 1980's)

Joan Hare
Assistant Research Scientist
Institute of Molecular Biophysics
Florida State University
Tallahassee, Florida

Dear Shirley,

It has been a real privilege to be the hub of all the correspondence among people who respected, admired, and loved Herbert Taylor. Because of it I have developed in my mind a very strong picture of who Herb Taylor was, and the striking things that have come through to me are that we all are describing the same elements of him. How wonderful that he showed us all, in one way or another, the same man. When I mention that look in his eye when he was entertaining new ideas, every-

Taylor and Joan Hare, his thirty-first Ph.D. student, relax following the Ph.D. hooding ceremony, August 1984

body says, yeah I knew that. When Russell Stevens talked about his wonderful way of laughing at himself, we all can remember times like that. When Terry Meyers told his story of booking Herb into a distant hotel at the ASCB meeting the year Herb took over the presidency, we can all imagine the HerbTaylor we knew taking it in stride with his usual sense of humility and laughing at the irony of the situation, walking the extra mile to the meeting site. When Dan Riggs told his fish story, I remembered that moment and recognized the man with the sometimes devilish sense of humor. When I talked with John Hozier, we each knew the dread of Herb's willingness to entertain "crazy ideas." We decided probably all of his students had worked on at least one wild idea.

Jesse Sisken mentioned his tolerance, and I think enjoyment, of students. I am reminded of the time when the gerbils Marcia was keeping in a cage by her desk, which Herb didn't entirely approve of, escaped, and we didn't know where they had gone until some weeks later when we began to see them scurrying across the floor late at night. One late afternoon when we were returning from Biology Colloqium we again saw them running, this time more boldly than usual between different rooms in the lab. In the confidence that only a "mob" can engender, we decided to catch them. Of course, they were much faster than us, and knew more places to hide than we could imagine. Being graduate students, however, and because it was by then after hours by a few minutes, we began to really get into it, pulling out a centrifuge, crawling on the floor under tables, waving a broom around after the poor things, hollering, "there he goes!" In the midst of this pandemonium, Dr. Taylor walked in, just back from talking with the colloquium speaker, to see most of his graduate students acting like out-of-control 10 year olds (to be generous). What do you think? Did he get mad, tell us to put the lab back in order, call in the exterminator? Nope, he asked "where did you see it last?" and joined in the fun! I don't remember whether we caught one or not; that would certainly have been anticlimactic, anyway.

Like the gerbil hunting, to get the most from a Herb Taylor lab experience required the ability to see things differently, to be willing to find and use whatever tools were needed to help you uncover more of the story. Just because a tool or device or chemical wasn't available was no reason to give up the experiment. Make your own methylase, make your own cell line, make your own light exposure table, design your own frog restrainer (Herb's own creation). But Herb gave you the idea that if you could imagine it, you could build it (with a little help). Well at least you could try. We did seem to have a lot of prototypes of various devices lying around the lab, some from earlier eras that I hadn't the faintest ideas what to do with when I later inherited all the lab stuff. Of course there were homemade grasshopper cages from the early years in Florida (I discovered what they were this year as I explored the old papers). I think we in Herb Taylor's lab probably could have kept our machine and carpentry shops busy full time, just by ourselves. We did almost everything from scratch. It did take longer, and as the newer techniques and methods of molecular biology came out, degrees took longer and longer to finish. But, by gosh, when you got a degree out of Herb Taylor's lab, you darn sure knew how to do it all yourself. We've all seen the products of the production-line labs where one person does only a part of the project and how helpless they are on their own. And patience; boy did we learn patience—to pick it up and try again until it worked.

Many of us have talked about his devotion to discovery. With Herb Taylor I developed a real love for designing experiments. What could be more fun than sitting with him and working out a plan; finding a clever way to do something that you hadn't been able to do before. Herb always gave us a sense of the ongoing nature of scientific investigation. With him I think we all learned what science really is, and we were infected with the itch to discover new understandings of the processes of life. I found myself somewhat shocked later to discover not every scientist has that fundamental drive to know. His was a very special love of discovery—untainted by so many of the things that drive other people: power, fame, prestige, money, and self-interest. He was willing to wait for the best and to believe in it.

These qualities I have inherited from Herb Taylor, and for these I am extremely grateful—the love of discovery, devotion to the search, excitement in the design, the ability to make it yourself and fix it yourself, and patience. These have served me well in the life that followed graduate school.

The other realization about Herb that came as I looked through old papers, and talked to and read what earlier-era colleagues had to say, was that Herb was fascinated by the same scientific puzzles all his life. You remember the long talks in his office and the incredible number of relevant experiments from others he could recall? All those things are there in his earlier work too. You can see the beginning of an idea and its development years later. What a keen intellect! How much we all admired his ability to put things together.

He taught us not only science, but real life as well; a camping or adventuring trip was always an approved excuse for absence. He taught us how to deal gently with people. He taught us how to write; he was an incredibly good editor. He could cut to the heart of it very quickly. I still see his elegant handwriting in the margins of my papers and ask myself, how would Herb fix this? And wouldn't Herb love this experiment?

Joan Hare
(Ph.D., FSU, 1984; postdoctoral research scientist, 1985–1987)

Clare Hasenkampf

Assistant Professor
University of Toronto at Scarborough
Scarborough, Ontario, Canada

Hi Joan,
Sorry I've been so long responding to your last e-mail. I put it aside

to try to see if I could recall something for the memorial book. I cannot really recall an individual story of Dr. Taylor. What I remember the best (probably because it is most recent) is from his visits here in Toronto. We started talking a bit about science, and I was asking him for any reports on evidence that homologous chromosome regions are in proximity to each other when they replicate. He began scrolling through his memory and came up with an amazing amount of information from the recent to articles from 20-plus years ago. He still seemed so enthusiastic about understanding chromosomes.

The other thing I'll always remember is when I submitted my Ph.D. work to Dr. Taylor for consideration for publication in *Chromosoma*. He said he had never seen anyone get so much information out of so little data! (I am still not sure it was a compliment! … but he did publish it.)

I don't know what the time frame is for articles for the Taylor *Chromosoma* edition, but I am working frantically on a project that might be appropriate for *Chromosoma*. Whether or not I can get it done in time is another matter.

Clare Hasenkampf
(Ph.D., with Margaret Menzel, FSU, 1983; lab rotation student with JHT)

Karin Sturm
Basel, Switzerland

Dear Joan,

Science, DNA replication, methylation, fishing, finding the perfect spot for blueberry-picking in the woods around Tallahassee, bringing unknown plants into the lab, having foreign students in the lab, interest in foreign cultures, music, gorillas—all a basic and general curiosity and appreciation of all aspects of life and nature. Fishing and replication were of equal importance—each in its time.

Herb seemed to be driven not by "the right career move" in his research but by his own persistence in trying to figure out a problem. Maybe that was a luxury of that time, maybe he was at a point in his career where he didn't need that anymore—full professor, member of the National Academy, etc., but I haven't met anybody since with such calm determination. When I think of him, the picture of a reflective, appreciative person comes to mind, not that of a successful high-power scientist.

He always saw the good, interesting sides in people. One important time for me was when I took a "sabbatical" one year into my Ph.D. time to go to a retreat center in the Cascades—trying to find out whether science was the right thing for me. He was the last person I told about my plans and worries; I remember how much I dreaded that conversation—and he just listened and said that everybody has to go and sit on top of a mountain every once in a while and think. "Why didn't I talk to him earlier??" In fact at the time he was thinking about how to retire. He told me, I'd always have a place in his lab should I decide to stay in science—I did come back and am still a scientist. Certain that I want to and can do science but not so certain on how to make a livable career out of doing basic research.

He was genuine and caring—even paid the students one summer out of his own pocket when he was between grants. (I think Marcia was still working on methylation then, 1980 or '81, I'm not sure.) He's asked key questions (although that wasn't obvious to me at the time!), questions not only about the mechanism of replication, but also about replication as a way of patterning a cell in differentiation or development—topics that are quite "in" now. What makes some origins of replication active and others not? How is the replication process organized? Relative timing of loci on homologous chromosomes—early and synchronous replication of potentially active regions in the genome, late and asynchronous replication of inactive regions such as the inactive X-chromosome in female cells (Barbie's work).

A couple of thoughts keep coming up. One is the persistence, or perhaps perseverance, he had in continuing his research even under adverse circumstances. While we were washing Pasteur pipettes and "recycling paper towels," Christoph (who was at the Basel Institute of Immunology) already used all disposables, kits. He didn't pour a bacterial plate himself until he was a postdoctorate in San Francisco. The questions Herb was interested in were so important for him that it was worth it to struggle through these times and to pull a bunch of students along.

He was always for the student—he would give anybody a chance if he or she were interested. If we were ready to listen, he was there to tell. But he didn't prepare a menu for us on a silver platter, a Ph.D. in a defined time, with the project already laid out, just to fill in the blanks in an established system—not so often done these days. He was very positive about research; negative aspects are part of normal life, but he never complained about things in the way some of our professors dealt with frustrations. That's something very important for us right now, too—to figure out what we want and then pursue those questions, change focus or maybe even direction along the way, apply new knowledge as you said, but believe in our own judgment, do science the right way, ask critical simple questions. Good experiments will always stand on their own.

I remember one of his lectures on polymerase-a during which all of sudden he stopped, still facing the blackboard and mumbling to himself. He was trying to figure out how what he had just said fit with an observation he had made earlier about fragment size. At first I thought I had lost the thread because my English wasn't so good then, and I just simply didn't understand. But he was just always moving those questions around in his head, looking at them from different aspects.

Karin Strum
(Ph.D., FSU, 1986)

Dan Riggs
Associate Professor
University of Toronto
Toronto, Ontario, Canada

Dear Shirley,

We were very saddened to hear about Herb. When I was a student in his lab, I did not really appreciate all that he was, but with the benefit of 15 years of aging, I now realize how fortunate I was to have him accept me. I feel very privileged to have known such a kind, gentle man, one who was a deep thinker that most people held in awe. As my career develops, I am cognizant of the politics of science and the enormous egos of many scientists. While I am sure that Herb felt himself above most other scientists, he never displayed it and was always the consummate professional. During my tenure as a student I grew more and more fond of him, and I was determined to succeed and make him proud of me. I asked him some years ago to put a letter of recommendation on file for me at FSU, and I am quite proud of his assessment of me as a person and a scientist. Often when things are going poorly I recall some of these words and his advice on various things, and I take comfort in them and the strength and resolve they impart to me. As a teacher, I particularly enjoy telling my students about his landmark experiments. Their elegant simplicity makes them comprehensible to the novice, and I enjoy the reactions I get from students, which range from "cool" to "that guy was really smart." Now I shall treasure them all the more.

Upon the passing of a loved one or one well respected, introspection reveals to us our qualities and our flaws and perhaps our own mortality. We often wish that we had been closer or nicer or better correspondents, etc. Since I left Tallahassee, there have been many times when I wished that I could be teleported into a canoe with him to ask his advice on a seemingly insurmountable problem. I tried to imagine what he would say, and this often helped. I have recalled many memories of my time in Tallahassee in the past three weeks and what Herb meant to me. I aspire to be like him and hope that in my passing, people will think of me in the same way that I think of him. In many ways he made a difference to those around him, and his impact on science cannot be measured.

I am unsure if you have ever heard this story or not, and as my letter so far has been a rambling sentimental mess, I will close with a more lighthearted reminiscence. I don't believe I knew any graduate students who didn't view Herb's intellect as something quite unnatural. Seminar presenters often did not care to have him attend, for he would always ask a probing question that, quite often, the speaker had never considered and/or was not prepared to answer. Herb always asked the question in a good natured and friendly fashion, which made taking one's medicine somewhat more palatable. Although we hated to be embarrassed in this way, deep down we appreciated that it made us work harder and consider things more carefully. Of course, he did not reserve his questions for in-house speakers, and we took some pleasure in imagining colloquium speakers behind a podium with a target upon it. One of my most vivid memories of Herb centers on a visit by a young marine biologist. In an interesting seminar, he presented information on light quality versus depth, species marking, and pigmentation and how these factors influenced feeding and predation. I was often bored or overwhelmed by most colloquia, but this one was interesting, the man was a good speaker, and he seemed to have a natural curiosity and enthusiasm for his subject that most other speakers lacked. Weighing this, I began to feel sorry for the speaker as the question period approached. It was Herb's turn. First, as usual, he set the stage by asking for a clarification, and then about the spectral qualities at around 20 feet down. I thought this was probably heading for some esoteric biophysics, but then he literally "set the hook," asking, "Taking all of these aspects into consideration, could you tell me what type of lure I should use to catch a 10-pound bass?" This of course brought the house down, wrapped up the seminar, and gave us something to laugh about until this day and beyond.

I regret that we will not be able to come to the memorial service. We hope that you are doing well and will take comfort from your family and the many friends who will attend. We very much enjoyed your visit with us several years ago and wish for you to know that you have an open invitation here. We cannot thank you enough for introducing us to the McMichael Art Gallery and the Group of Seven. We have fallen in love with their style and, as we discover more and more of Canada and its rugged beauty, we grow to love their art even more. Thank you for being a part of our lives, and please come to visit us again.

Sincerely yours,
Dan Riggs
(Ph.D., FSU, 1986)

Taylor's catch on Lake Wabatongushi, wilderness camp, Northwestern Ontario, 1997

José V. Lopez, Ph.D.
Harbor Branch Oceanographic Institution
Fort Pierce, Florida

  I essentially began my scientific career as a master's student in Dr. Taylor's laboratory. I was a fairly raw, untested student with a less than distinguished undergraduate record in biology, but Herb decided to give me a break anyway with a research assistantship in his lab when I first arrived at FSU. I was grateful then and now that he decided to give me that chance, and I tried not to let him down. During those few years at FSU, he gave me the freedom to discover and learn through my mistakes. Herb was an excellent role model. He taught me the proper conduct of science, and proper conduct, through his example and his writings. I will never forget his patience and legacy and the goodness that he and Shirley shared with everyone they met.

José Lopez
(M.S., FSU, 1988)

CLOSING THE CIRCLE

Within the circle of our lives
we dance the circle of the years,
the circles of the seasons
within the circles of the years,
the cycles of the moon
the circles of the seasons,
the circles of our reasons
within the cycles of the moon.

Again, again we come and go,
changed, changing. hands
join, unjoin in love and fear,
grief and joy. The circles turn,
each giving into each, into all.
Only music keeps us here,

each by all the others held.
In the hold of hands and eyes
we turn in pairs, that joining
joining each to all again.

And then we turn aside, alone,
out of the sunlight gone

into the darker circles of return.

"Song (4)," Wendell Barry

Song (4) from COLLECTED POEMS: 1957-1982 by Wendell Berry. Copyright © 1985 by Wendell Berry. Reprinted by permission of North Point Press, a division of Farrar, Straus and Giroux, LLC.

List of friends, colleagues, students, and postdoctoral fellows of Dr. J. Herbert Taylor

Abele, Lawrence — colleague; Chair, Department of Biological Science, 1983–1991
Provost, Florida State University, Tallahassee, Florida

Adegoke, Joseph A. — Ph.D., Taylor lab, FSU, 1974;
Department of Genetics, Lund University, Lund, Sweden, and Department of Zoology, Obafemi Awolowo University , Ille-Ife, Nigeria

Alberts, Bruce — colleague
President, National Academy of Sciences, Washington, D.C.

Angeloni, Stephen — B.S., Taylor lab, FSU; M.S., Roberts lab, FSU; Ph.D., Virginia Polytechnic Institute
Center for Vaccine Development, School of Medicine, University of Maryland, Baltimore, Maryland

Ashley, Terry — graduate student, Taylor lab, FSU, 1965–1966; Ph.D., M. Menzel lab, FSU, 1970
Associate Research Professor, Department of Genetics, School of Medicine, Yale University, New Haven, Connecticut

Badr (Mourad), Effat A. — Ph.D., Taylor lab, Columbia., 1962
Alexandria, Egypt; retired as Chair, Department of Biology, University of Alexandria

Bass, Hank W. — Geneticist, JHT Professorship selection committee
Department of Biological Science, Florida State University, Tallahassee, Florida

Beerman, Wolfgang (deceased) — visiting scientist
Formerly Director, Max Planck-Institut für Biologie, Tübingen, Germany

Berg, Claire (deceased), — Ph.D., Taylor lab, Columbia, 1966
Formerly University of Connecticut, Storrs, Connecticut

Boggs, Barbara — B.S., Taylor lab, FSU, 1980; Ph.D., Baylor University, 1990
Assistant Research Professor, Department of Molecular and Human Genetics, Baylor College of Medicine, Houston, Texas

Boyes, Barbara — M.S., Taylor lab, Columbia, 1964; FSU, summer 1964–65
Ottawa, Ontario, Canada; retired as Federal Government Research Scientist in Genetic Toxicology

Burnham, Maria (Piraino) — Ph.D., Taylor lab, Columbia, 1969
Essex, Massachusetts; Department of Biology, University of Massachusetts, Boston, Massachusetts

Chambers, Jasemine (Choy) — Ph.D., Taylor lab, FSU, 1982; J.D., George Washington University Law School, 2001
Director, Technology Center, U.S. Patent and Trademark Office, Alexandria, Virginia

Clark, Roger — postdoctoral fellow, Taylor lab, 1971–1972
Department of Biology, University of Denver, Denver, Colorado

Crouse, Helen — visiting scientist, Taylor lab, 1970–1978
Hayesville, North Carolina; retired Drosophila geneticist

Cunningham Peoples, Hervey — M.S., Taylor lab, FSU, 1970
Scientific author, Longboat Key, Florida; Past positions: medical research biologist; high tech marketer

De, Deepesh N. — Ph.D., Taylor lab, Columbia, 1960
Kalyani, India; Professor emeritus, Applied Botany, Agricultural and Food Engineering Department, Indian Institute of Technology

Easton, Dexter M. — fellow Unitarian Church member, friend
Department of Biological Science, Florida State University, Tallahassee, Florida

Erickson, Leonard C. — Ph.D., Taylor lab, FSU, 1974
Robert Wallace Miller Professor of Oncology, Director of Basic Research, Indiana University Cancer Center, Indianapolis, Indiana

Evenson, Donald P. — postdoctoral fellow, Taylor lab, FSU, 1970
Brookings, South Dakota; retired from South Dakota State University

Gall, Joseph G. — colleague, JHT Professorship selection committee
Carnegie Institution, Baltimore, Maryland

Gaulden, Mary Esther — Blandy Farm classmate at University of Virginia and longtime friend
Adjunct Professor, Department of Radiology, Southwestern Medical Center, University of Texas, Dallas, Texas

Gerbi, Susan — undergraduate student, Taylor genetics class, Columbia, 1964; Ph.D., Yale, 1970
Professor and Chair, Department of Molecular Biology, Cell Biology, Biochemistry, Division of Biology and Medicine, Brown University, Providence, Rhode Island

Herb Taylor and sixteen of his former graduate students who returned to Tallahassee, Florida in May of 1990 to wish him a fond farewell and a happy retirement.

Grossman, Mickey — M.S. 1978; graduate student, Taylor lab, FSU, 1978–1984
Talent agent, Linda Jack Agency, Chicago, Illinois

Guy, Arthur — Ph.D., Taylor lab, FSU, 1977
Federal Bureau of Investigation

Hare, Joan — Ph.D., Taylor lab, FSU, 1984; postdoctoral fellow 1985–1987
Research Scientist, Florida State University, Tallahassee, Florida

Harris, Marcia (Applegate) — Ph.D., Taylor lab, FSU, 1984; M.D., 1992
Physician, Family Medicine/Obstetrics, Floyd, Virginia

Hasenkampf, Clare — graduate student, Taylor lab, 1981; Ph.D., M. Menzel lab, FSU, 1982
Associate Professor of Botany, Division of Life Sciences, University of Toronto at Scarborough, Scarborough, Ontario, Canada

Haut, William F. — Ph.D., Taylor lab, Columbia, 1966; FSU, 1964–1965
Sierra Vista, Arizona; retired from Biology Department, Hofstra University

Hennig, Wolfgang – colleague
Professor of Genetics, Department of Physiological Chemistry, and Pathobiochemistry, Johannes Gutenberg-University Mainz, Mainz, Germany

Henry, Dan — B.S., Taylor lab, FSU; doctorate in Dentistry
Private dental practice, Pensacola, Florida

Hershey, Howard — postdoctoral fellow, Taylor lab
Department of Biology, Indiana University, Bloomington, Indiana

Hofer, Kurt G. — colleague, JHT Professorship selection committee
Tallahassee, Florida; retired 2003 from Institute of Molecular Biophysics, Florida State University

Hopkins, Bob — friend
Sierra Vista, Arizona; retired science teacher

Hozier, John C. — Ph.D., Taylor lab, FSU, 1975
Department of Pathology, University of New Mexico, Albuquerque, New Mexico

Hurst, Myrna — Taylor's secretary, Institute of Molecular Biophysics, FSU, 1973–1985
Tallahassee, Florida; retired from Florida State University

Jona, Roberto — postdoctoral fellow, Taylor lab
Dipartimento di Coltivazioni Arboree, Universita di Torino, Torino, Italy

Kasha, Michael — colleague; established Institute of Molecular Biophysics, FSU, 1960; Director IMB, FSU, 1960–1975
Tallahassee, Florida; retired from Department of Chemistry, Florida State University

Kosan, Joan Dalheim — Ph.D., Taylor lab, Columbia, 1959
New York, New York; retired from Department of Biological Sciences, New York City College of Technology

Kurek, Michael Paul — Ph.D., Taylor lab, FSU, 1976
Principal Consultant, Biotechnology Business Consultants, Ann Arbor, Michigan

Lark, Gordon & Cynthia — colleagues
Salt Lake City, Utah; retired from Department of Biology, University of Utah

Laughlin, Tom — Ph.D., Taylor lab, FSU, 1979
President, Mist-On Systems Inc., Grapevine, Texas

Lee, Jeanne Tung — M.S., Taylor lab, Columbia, 1959
last known address: Kowloon, Hong Kong

Lopez, José — M.S., Taylor lab, FSU, 1987; Ph.D., George Mason University, 1995
Division of Biomedical Research, Harbor Branch Oceanographic Institution, Fort Pierce, Florida

McGraw, Royal Alfred — Ph.D., Taylor lab, FSU, 1982
College of Veterinary Medicine, University of Georgia, Athens, Georgia

McMaster, Rachel (deceased) — postdoctoral fellow, Taylor lab, Columbia, 1963
Formerly University of Rochester, Rochester, New York

Mego (formerly Straubing), Nancy — M.S., Taylor lab, FSU, 1969
Owner/Architectural Artist, Architectural Graphics, Napierville, Illinois; Formerly at Papanicholaou Cancer Research Institute, University of Miami

Mego, William A. — Ph.D., Taylor lab, FSU, 1970
Biotechnology Coordinator, Consultant, Napierville, Illinois; Formerly Department of Biological Sciences, University of Chicago

Miller, Oscar — colleague
Charlottesville, Virginia; retired as Chair, Department of Biology, University of Virginia

Miner, Philip (deceased) — Ph.D., Taylor lab, Columbia, 1967; FSU 1964–1966

Mitra, Sandhya (Ghosh) — Ph.D., Taylor lab, Columbia, 1955
New Delhi, India; retired from Birla Institue of Technology and Science

Moerland, Timothy — Chair, JHT Professorship selection committee
Professor and Chair, Department of Biological Sciences, Florida State University, Tallahassee, Florida

Moore, Eleanor — fellow Sierra Club member, friend
Tallahassee, Florida, retired social worker

Moses, Montrose — Brookhaven colleague
Durham, North Carolina; retired from Department of Cell Biology, Medical School, Duke University

Myers, Terry L. — Ph.D., Taylor lab, FSU, 1969; M.D., University of Virginia
Plano, Texas; retired as Professor and Chair of Pediatrics and Associate Dean of Clinical Affairs, School of Medicine, Texas Tech University

Nicklas, Bruce — colleague
Professor and Chair, Department of Cell Biology, Duke University, Durham, North Carolina

Pelling, Claus — visiting scientist, colleague
last known address: Max Planck-Institut für Biologie, Tübingen, Germany

Petersen, Arnold (deceased) — postdoctoral fellow, Taylor lab, Columbia, 1962

Pollock, Jim — postdoctoral fellow, Taylor lab, 1976–1979
Florida Department of Law Enforcement, Florida Crime Lab, Division of Serology, Jacksonville, Florida

Rho, Young Chu — M.S., Taylor lab, 1969
last known address: Department of Zoology, College of Natural Sciences, Seoul National University, Seoul, Korea

Riggs, C. Daniel — Ph.D., Taylor lab, FSU, 1986
Associate Professor, Department of Botany, University of Toronto at Scarborough, Scarborough, Ontario, Canada

Roeder, Martin — colleague
Tallahassee, Florida; retired from Department of Biological Science, Florida State University,

Rosenberg, Marvin — Ph.D., Taylor lab, Columbia, 1966
Fullerton, California; Professor emeritus of Biology and Associate Dean, College of Natural Science and Mathematics, California State University, Fullerton

Santiago, Milagros (Nieves) — B.S., Taylor lab, FSU, 1981
Pharmaceutical scientist; last known address: Hilltop Research, St. Petersburg, Florida

Schwartz, Alice Adams — Ph.D., Taylor lab, FSU, 1974
last known address: Pfafftown, North Carolina

Shah, Vinod C. — Ph.D., Taylor lab, Columbia, 1960; later a visiting scientist in the Taylor lab, FSU
Ahmedabad, Gujarat, India; retired as Dean of the School of Sciences, Gujarat University

Shandl, Emil K. — Ph.D., Taylor lab, FSU, 1970
last known address: Hollywood, Florida

Shull, Ken — graduate student, Taylor lab, FSU; Ph.D., M. Menzel lab, FSU, 1973
Professor, Department of Biology, Appalachian State University, Boone, North Carolina

Sisken, Jesse E. — Ph.D., Taylor lab, Columbia, 1957
Lexington, Kentucky; retired as Professor, Department of Microbiology and Immunology, Chandler Medical Center, University of Kentucky

Stevens, Russell (deceased) — University of Tennessee colleague
National Research Council, Washington, D.C.

Sturm, Karin (Berger) — Ph.D., Taylor lab, FSU, 1986
Family manager/community activist, Grossh ö chstetten, Switzerland; Formerly at Children's Medical Research Foundation, Westmead, NSW, Australia

Travis, Joseph — colleague, Department of Biological Science, FSU
Director, School of Computational Science and Information Technology, Florida State University, Tallahassee, Florida

Van Wart, Harold — colleague; former IMB professor
Director Biochemistry and Molecular Biology, Roche Pharmaceuticals, Palo Alto, California

Wall, Lynne — postdoctoral fellow, Taylor lab, 1983–1984
Strategic Resources Manager and Research Director Human Sciences, United Kingdom Ministry of Defence, Overton, Hampshire, England

Watanabe, Shinichi — Ph.D., Taylor lab, FSU, 1980
Abbot Laboratory, Abbot Park, Illinois

Winchester, John W. — fellow Sierra Club member, friend
Tallahassee, Florida; retired from Department of Oceanography, Florida State University

Wise, Dwayne — graduate student, Taylor lab; Ph.D., M. Menzel lab, FSU, 1972
Professor, Department of Biological Sciences, Mississippi State University, Starkville, Mississippi

Wolff, Sheldon — colleague
Mill Valley, California; Professor emeritus, University of California

Wu, Madeline C. and Jung-Rung — postdoctoral fellows
Chair, Department of Biology, Hong Kong University of Science and Technology, Kowloon, Hong Kong

List of Taylor Family

Foster, June (Hoover), sister-in-law
M.A., Art Education, University of Maryland, Chestertown, Maryland; retired as Art Teacher, Maryland Public Schools

Ireland, Lynne (Taylor), daughter
M.A., Early Childhood Education, University of Florida; Director, Boopa Werem Aboriginal Preschool, Cairns, Queensland, Australia

Taylor, Lucy, daughter
B.S., Art, FSU; Ceramics Teacher, Agoura Hills High School, Thousand Oaks, California

Taylor, Michael, son
M.S., Optical Engineering, University of Florida; Vice-president for Strategic Planning, netMercury, Williamstown, Massachusetts

Taylor, Shirley (Hoover), married J. Herbert Taylor in 1946
M.A. and Ph.D. in Biology, University of Virginia, Taylor lab manager, FSU ; Williamstown, Massachusetts; retired as Sierra Club's national leader on coastal and marine policy issues

Taylor, Tom Dotson, brother
Bodega, California; retired as Education director, Western Civil Defense Region

J. HERBERT TAYLOR

Curriculum Vitae

Department of Biological Science and Institute of Molecular Biophysics, Florida State University, Tallahassee, FL 32306-4380. Biology. b. Corsicana, Texas, January 14, 1916; m. Shirley Hoover, 1946; 3 children. B.S. (Biology and Math) Southeastern Oklahoma State, 1939; M.S. (Botany and Bacteriology), University of Oklahoma, 1941; Ph.D. (Biology), University of Virginia, 1944. U.S. Army 1943–1946 (South Pacific). Asst. Professor, Plant Science, University of Oklahoma, 1946–1947; Assoc. Professor of Botany, University of Tennessee, 1947–1951; Assoc. Professor of Botany, Columbia University, 1951–1954; Assoc. Professor of Botany, 1954–1958; Professor of Cell Biology, Department of Zoology and Botany, 1958–1964; Professor of Biological Science, Institute of Molecular Biophysics, Florida State University, 1964–1980; Professor Emeritus, 1990–1998. Associate Director, Institute of Molecular Biophysics, 1970–1980; Director, Institute of Molecular Biophysics, 1980–1985.

Consultant in Biology, Oak Ridge National Laboratory, 1948–1951; Research Collaborator, Brookhaven National Laboratory, 1952–1958; Genetics Panel, NSF, 1963–1966; Genetics Training Committee, NIH, 1966–1970; Consultant, NIH Genetic Training Program, 1970–1973.

Honors, Offices, and Lectureships: Robert O. Lawton Distinguished Professor (1983). Elected to National Academy of Sciences (1977). Citation by Board of Regents for outstanding research and scientific accomplishments (1977). Commendation by a special resolution of the Senate and House of Representatives for outstanding achievements in research and teaching while employed at the Florida State University (1977). Award for Meritorious Research from Michigan State University (1960). Guggenheim Fellow at the California Institute of Technology (1958–59). Jefferson Award for Ph.D. research at University of Virginia (1944). Blandy Farm Fellowship at the University of Virginia (1941–1943). University Scholarship at the University of Oklahoma (1940–41).

American Society of Cell Biology, Council (1964–1970); President Elect (Program Chairman), 1968; President (1969); Past President (1970). Co-chairman of the Gordon Research Conference: Cell Structure and Metabolism (1965).

Sigma Xi National Lectureship (1960).

Society Memberships: Biophysics Society (charter member), American Society of Cell Biology (charter member), Radiation Research Society (charter member), Genetic Society of America (1948–1998), Botanical Society of America (1950–1976).

Editorial Positions: Editorial Board of Journal of Cell Biology (1963–1966); Editorial Board of Genetics (1966–1971); Regional Editor of Chromosoma (1966–1993), Nucleus (1958–1990), and Mutation Research (1967–1977); Editorial Board of BioScience (1973–1976), Regional Editorial Board of Environmental and Experimental Botany (1961–1990).

Major Research Accomplishments: (1) Demonstrated the labeling with ^{3}H-thymidine of the two DNA subunits of chromosomes and their semiconservative segregation during cell division (1956–57); first evidence for semiconservative replication of DNA. (2) Found the opposite polarity of the two subunits of chromosomes, indicating their analogy to the two chains of the Watson-Crick helix (1958–59). (3) Demonstrated the pattern of labeling of chromosomes over the cell cycle, including the discovery of the late-replicating X chromosome in mammals (1960). (4) Demonstrated that physical exchanges between homologous chromosomes occur during meiosis (1965). (5) Found that replication of DNA in mammalian chromosomes occurs by the production of small segments and later showed that some segments contain ribonucleotides (1969–1973). (6) Isolated and partially characterized the replication complex from mammalian cells in culture (1973–1977). (7) Obtained evidence concerning the kinetics and regulation of initiation in highly synchronized cells at the beginning of S phase. Showed that potential sites are available at 4-micron intervals along the DNA but that, in the normal cycle of cultured fibroblasts, only one in

15–20 of the potential origins is actually used in initiation of DNA replication (1973–1977). (8) (In collaboration with Shinichi Watanabe, graduate student) Cloned many EcoRI segments of Xenopus DNA into an Escherichia coli plasmid and tested these for origins of replication by injecting the supercoiled recombinant plasmids into unfertilized Xenopus eggs. Some sequences function as origins. A 505-base-pair segment containing an origin was sequenced, and experiments were continued to determine the sequences at functional origins (1980–1986). (9) (In collaboration with Karin Sturm and R. Alfred McGraw) Demonstrated by sequencing DNA that there are differences in modification (methylation) of satellite DNA's in cells of various differentiated bovine tissues. Sperm and chorion are methylated at few sites compared to thymus, brain, liver, and kidney cells (1981–1986). In collaboration with Joan Hare) Demonstrated that methylation of cytosine affects mismatch repair in mammalian cells (1982-1990).

BIBLIOGRAPHY

1. Eigsti, O. J., and H. Taylor. 1941. *The induction of polyploidy in phlox by colchicine.* Proc. Okla. Acad. Sci., *120–122.*

2. Taylor, H. 1941. *A physiological study of diploid and related tetraploid plants.* Proc. Okla. Acad. Sci., *137–138.*

3. Taylor, H. 1945. *Cyto-taxonomy and phylogeny of the Oleaceae.* Brittonia 5: *337–367.*

4. White, O. E., J. H. Taylor, and B. M. Speese. 1946. *Begonia species hybrids.* J. Hered. *37: 67–70.*

5. Stevens, R. B., and J. H. Taylor. 1948. *Photomicrography at your convenience.* Science *108: 420–421.*

6. Taylor, J. H. 1949. *Chromosomes from cultures of excised anthers.* J. Hered. *40: 87–88.*

7. Taylor, J. H. 1949. *Increase in bivalent interlocking and its bearing on the chiasma hypothesis of metaphase pairing.* J. Hered. *40: 65–59.*

8. Taylor, J. H. 1950. *The duration of differentiation in excised anthers.* Am. J. Bot. *37: 137–143.*

9. Taylor, J. H. 1950. *Cytotaxonomy. In* Families of Dicotyledons. A. A. Gunderson, ed. Chronica Botanica Co., Waltham, Mass. pp. *20–24.*

10. Taylor, J. H. 1953. Autoradiographic detection of incorporation of P^{32} into chromosomes during meiosis and mitosis. Exp. Cell Res. 4: 164–173.

11. Taylor, J. H. 1953. Intracellular localization of labeled nucleic acid determined with autoradiographs. Science 118: 555–557.

12. Taylor, J. H., and S. H. Taylor. 1953. The autoradiograph—a tool for cytogeneticists. J. Hered. 44: 129–132.

13. Taylor, J. H., and R. D. McMaster. 1954. Autoradiographic and microphotometric studies of desoxyribose nucleic acid during microgametogenesis in Lilium longiflorum. Chromosoma 6: 489–521.

14. Moses, J. J., and J. H. Taylor. 1955. Desoxypentose nucleic acid synthesis during microsporogenesis in Tradescantia. Exp. Cell Res. 9: 474–488.

15. Taylor, J. H., R. M. McMaster, and M. J. Caluya. 1955. Autoradiographic study of incorporation of P^{32} into ribonucleic acid at the intracellular level. Exp. Cell Res. 9: 460–473.

16. Taylor, J. H. 1956. Autoradiography at the cellular level. In Physical Techniques in Biological Research. G. Oster and A. W. Pollister, eds. Academic Press, New York. pp. 545–576.

17. Taylor, J. H., P. S. Woods, and W. L. Hughes. 1957. The organization and duplication of chromosomes as revealed by autoradiographic studies using tritium-labeled thymidine. Proc. Natl. Acad. Sci. USA 43: 122–128.

18. Taylor, J. H. 1957. The time and mode of duplication of chromosomes. Am. Nat. 91: 209–221.

19. Taylor, J. H. 1958. Incorporation of phosphorus-32 into nucleic acids and proteins during microgametogenesis of Tulbaghia. Am. J. Bot. 45: 123–131.

20. Taylor, J. H. 1958. The mode of chromosome duplication in Crepis capillaris. Exp. Cell Res. 15: 350–357.

21. Taylor, J. H. 1958. Sister chromatid exchanges in tritium-labeled chromosomes. Genetics 43: 515–529.

22. Taylor, J. H. 1958. Duplication of chromosomes. Sci. Am. 198: 36–42.

23. McMaster-Kaye, R., and J. H. Taylor. 1958. Evidence for two metabolically distinct types of ribonucleic acid in chromatin and nucleoli. J. Biophys. Biochem. Cytol. 4: 5–11.

24. Taylor, J. H. 1958. Some uses of tritium in autoradiography. In Proceedings of the Symposium on Advances in Tracer Applications of Tritium. S. Rothchild, ed. The Sheldon Press, Boston. pp. 38–42.

25. Taylor, J. H. 1959. The organization and duplication of genetic material. Proc. X Int. Congr. Genet. I: 63–78.

26. Taylor, J. H. 1959. *Autoradiographic studies of the organization and mode of duplication of chromosomes. In* A Symposium on Molecular Biology. R. E. Zirkle, ed. University of Chicago Press, Chicago. pp. 304–320.

27. Taylor, J. H., and P. S. Woods. 1959. *In situ studies of polynucleotide synthesis in nucleolus and chromosomes. In* Subcellular Particles. American Physiological Society, Washington, D.C. pp. 172–185.

28. Woods, P. S., and J. H. Taylor. 1959. *Studies of ribonucleic acid metabolism with tritium-labeled cytidine.* Lab. Invest. 8: 309–318.

29. McMaster-Kaye, R., and J. H. Taylor. 1959. *The metabolism of chromosomal ribonucleic acid in Drosophila salivary glands and its relation to synthesis of deoxyribonucleic acid.* J. Biophys. Biochem. Cytol. 5: 461–467.

39. Taylor, J. H. 1959. *Autoradiographic studies of nucleic acids and proteins during meiosis in Lilium longiflorum.* Am. J. Bot. 46: 477–484.

40. Taylor, J. H. 1959. *Further studies on the mechanism of chromosome duplication. In* Proc. First Natl. Biophys. Conf. *Yale University Press, New Haven.* pp. 264–273.

41. Taylor, J. H. 1959. *High resolution autoradiography.* Okla. Conf. on Radioisotopes in Agriculture, *Atomic Energy Commission TID-7578.* pp. 69–75.

42. Taylor, J. H. 1959. *Autoradiographic studies of chromosome duplication and structure.* Okla. Conf. on Radioisotopes in Agriculture, *Atomic Energy Commission TID-7578.* pp. 123–129.

43. Taylor, J. H. 1960. *Autoradiography with tritium-labeled substances. In* Advances in Biological and Medical Physics, Vol. 7. *Academic Press, New York.* pp. 107–130.

44. Taylor, J. H. 1960. *Chromosome reproduction and the problem of coding and transmitting the genetic heritage.* Am. Sci. 48: 365–382.

45. Taylor, J. H. 1960. *Asynchronous duplication of chromosomes in cultured cells of Chinese hamster.* J. Biophys. Biochem. Cytol. 7: 455–464.

46. Taylor, J. H. 1960. *Nucleic acid synthesis in relation to the cell division cycle.* Ann. N.Y. Acad. Sci. 90: 409–421.

47. Taylor, J. H. 1960. *Duplication of chromosomes and related events in the cell cycle. In* Cell Physiology of Neoplasia. *University of Texas Press, Austin.* pp. 547–575.

48. Taylor, J. H. 1961. *Growth of the nucleus. In* Encyclopedia of Plant Physiology. *W. Ruhland, ed. Springer-Verlag, Heidelberg, Germany.* pp. 227–236.

49. Taylor, J. H. 1961. Physiology of mitosis and meiosis. Annu. Rev. Plant Physiol. *12: 327–344.*

50. Taylor, J. H. 1961. Control of DNA synthesis. *In* Conference on Molecular and Radiation Biology, *Nuclear Science Series Report NO. 31.* R. A. Deering, ed. National Academy of Sciences of the USA Publication 823, Washington D.C. pp. 12–20.

51. Taylor, J. H. 1962. Tritium and autoradiography in cell biology. *In* Tritium in the Physical and Biological Sciences, Vol. 2. Vienna, Austria. pp. 221–228.

52. Taylor, J. H., W. F. Haut, and J. Tung. 1962. Effects of fluorodeoxyuridine on DNA replication, chromosome breakage, and reunion. Proc. Natl. Acad. Sci. *48: 190–198.*

53. Taylor, J. H. 1962. Chromosome reproduction. Int. Rev. Cytol. *13: 39–72.*

54. Taylor, J. H. 1962. Chromosome reproduction and the problem of coding and transmitting the genetic heritage. *In* Science in Progress, 12th series. Yale University Press, New Haven and London. pp. 145–169. (A Sigma Xi-RESA Lecture for 1959–60.)

55. Morishima, A., M. M. Grumbach, and J. H. Taylor. 1962. Asynchronous duplication of human chromosomes and the origin of sex chromatin. Proc. Natl. Acad. Sci. USA *48: 756–763.*

56. Taylor, J. H. 1963. Control mechanisms for chromosome reproduction in the cell cycle. *In* Cell Growth and Cell Division. R. J. C. Harris, ed. Academic Press, New York. pp. 161–177.

57. Taylor, J. H. 1963. Effects of inhibitors of thymidylate synthetase on chromosome breakage and reunion. Exp. Cell Res., *Suppl. 9: 99–106.*

58. Grumbach, M. M., A. Morishima, and J. H. Taylor. 1963. Human sex chromosome abnormalities in relation to DNA replication and heterochromatinization. Proc. Natl. Acad. Sci. USA *49: 581–589.*

59. Taylor, J. H. 1963. DNA synthesis in relation to chromosome reproduction and the reunion of breaks. J. Cell Comp. Physiol. *Suppl. 1, 62: 73–86.*

60. Taylor, J. H. 1963. The replication and organization of DNA in chromosomes. *In* Molecular Genetics, Part I. *J. H. Taylor, ed. Academic Press, New York. pp. 65–111.*

61. Taylor, J. H., ed. *1963.* Molecular Genetics, Part I. *Academic Press, New York. 544 pp.*

62. Taylor, J. H. 1964. The arrangement of chromosomes in the mature sperm of grasshopper. J. Cell Biol. *21: 286–289.*

63. Taylor, J. H. 1964. *Regulation of DNA replication and variegation-type position effects. In* Symp. Int. Soc. Cell Biol., Vol. 3. R. J. C. Harris, ed. Academic Press, New York. pp. 173–190.

64. Taylor, J. H. 1965. *Distribution of tritium-labeled DNA among chromosomes during meiosis. I. Spermatogenesis in the grasshopper.* J. Cell Biol. *25: 57–67.*

65. Taylor, J. H., ed. 1965. Selected Papers on Molecular Genetics. *Academic Press, New York. 649 pp.*

66. Taylor, J. H. 1966. *The duplication of chromosomes. In* Probleme der Biologischen Reduplikation, *Vol. 3. Wissenschaftliche Konferenz der Gesellschaft Deutscher Naturforscher und Arzte, Semmering bei Wien 1965. P. Sitte, ed. Springer-Verlag, Heidelberg. pp. 9–28.*

67. Taylor, J. H. 1967. *Patterns and mechanisms of genetic recombination. In* Molecular Genetics, Part II. *J. H. Taylor, ed. Academic Press, New York. pp. 93–135.*

68. Taylor, J. H., ed. 1967. Molecular Genetics, Part II. *Academic Press, New York. 517 pp.*

69. Haut, W. F., and J. H. Taylor. 1967. *Studies of bromouracil deoxyriboside substitution in DNA of bean roots (Vicia faba).* J. Mol. Biol. *26: 389–401.*

70. Taylor, J. H. 1967. *The regulation of DNA replication in chromosomes of higher cells. In* Nucleic Acid Metabolism, Cell Differentiation and Cancer Growth. *E. V. Cowdry and S. Seno, eds. Pergamon Press, Elmsford, New York. pp. 231–239.*

71. Taylor, J. H. 1967. *Meiosis. In* Handb. d. Pflanzenphysiologie bd. XVIII. *W. Ruhland, ed. Springer-Verlag, Heidelberg, Germany. pp. 344–367.*

72. Taylor, J. H. 1968. *Structure and duplication of chromosomes. In* Genetic Organization. *E. W. Caspari, and A. W. Ravin, eds. Academic Press, New York. pp. 163–221.*

73. Taylor, J. H. 1968. *Rates of chain growth and units of replication in DNA of mammalian chromosomes.* J. Mol. Biol. *31: 579–594.*

74. Taylor, J. H., and P. Miner. 1968. *Units of DNA replication in mammalian chromosomes.* Cancer Res. *28: 1810–1814.*

75. Callan, H. G., and J. H. Taylor. 1968. *A radioautographic study of the time course of male meiosis in the newt Triturus vulgaris.* J. Cell Sci. *3: 615–626.*

76. Taylor, J. H. 1969. *Replication of organization of chromosomes.* Proc. XII Int. Congr. Genet. *3: 177–189.*

77. Schandl, E. K., and J. H. Taylor. 1969. *Early events in the replication and integration of DNA into mammalian chromosomes.* Biochem. Biophys. Res. Comm. *34: 291–300.*

78. Taylor, J. H. 1969. *Replication of chromosomal DNA and mechanisms of recombination.* In Genetics and Developmental Biology. H. J. Teas, ed. University of Kentucky Press, Lexington. pp. 150–164.

79. Taylor, J. H., W. A. Mego, and D. P. Evenson. 1970. *Structure and replication of eukaryotic chromosomes.* In The Neurosciences: 2nd Study Program. F. O. Schmitt, ed. Rockefeller University Press, New York. pp. 998–1013.

80. Taylor, J. H., T. L. Myers, and H. L. Cunningham. 1971. *Programmed synthesis of deoxyribonucleic acid during the cell cycle.* In Vitro 6: 309–321.

81. Schandl, E. K., and J. H. Taylor. 1971. *Oligodeoxyribonucleotides from pulse-labeled mammalian cells.* Biochim. Biophys. Acta 228: 595–609.

82. Evenson, D. P., W. A. Mego, and J. H. Taylor. 1972. *Subunits of chromosomal DNA. I. Electron microscopic analysis of partially denatured DNA.* Chromosoma 39: 225–235.

83. Taylor, J. H., A. G. Adams, and M. P. Kurek. 1973. *Replication of DNA in mammalian chromosomes. II. Kinetics of ^{3}H-thymidine incorporation and the isolation and partial characterization of labeled subunits at the growing point.* Chromosoma 41: 361–384.

84. Taylor, J. H. 1973. *Replication of DNA in mammalian chromosomes: isolation of replicating segments.* Proc. Natl. Acad. Sci. USA 70: 1083–1987.

85. Taylor, J. H. 1974. *Units of DNA replication in chromosomes of eukaryotes.* Int. Rev. Cytol. 37: 1–20.

86. Taylor, J. H., M. Wu, and L. C. Erickson. 1974. *Functional subunits of chromosomal DNA from higher eukaryotes.* Cold Spring Harbor Symp. Quant. Biol. 38: 225–231.

87. Hershey, H. V., and J. H. Taylor. 1974. *DNA replication in eucaryotic nuclei. Evidence suggesting a specific model of replication.* Exp. Cell Res. 85: 79–88.

88. Taylor, J. H. 1974. *Symposium No. 4: Meiosis.* Genetics 78 187–191.

89. Schwartz, A. G., and J. H. Taylor. 1974. *Repeated sequences in the DNA of the eukaryotic genome.* Chromosoma 49: 1–15.

90. Taylor, J. H. 1975. *The cell nucleus: session summary.* Mammalian Cells: Probes and Problems. Proc. 1st Los Alamos Life Sciences Symp., Los Alamos, N.M., 1973. Tech. Info Center, U.S. Research and Development Adm.

91. Hozier, J. C., and J. H. Taylor. 1975. *Length distributions of single-stranded DNA in Chinese hamster ovary cells.* J. Mol. Biol. 93: 181–201.

92. Hershey, H. W., and J. H. Taylor. 1975. *DNA replication in isolated nuclei. The fate of pulse-labelled DNA subunits.* Exp. Cell Res. 94: 339–350.

93. Taylor, J. H., M. Wu, L. C. Erickson, and M. P. Kurek. 1975. Replication of DNA in mammalian chromosomes. III. Size and RNA content of Okazaki fragments. Chromosoma 53: 175–189.

94. Taylor, J. H. 1976. Structural and functional subunits of chromosomes. In Adv. in Pathobiology 3: Developmental Genetics. C. M. Fenoglio, R. Goodman, and D. W. King, eds. Stratton Intercont. Med. Books Corp., New York. pp. 19–26.

95. Taylor, J. H., and J. C. Hozier. 1976. Evidence for a four micron replication unit in CHO cells. Chromosoma 57: 341–350.

96. Adegoke, J. A., and J. H. Taylor. 1976. Sequence programming of DNA replication over the S phase. Exp. Cell Res. 104: 47–54.

97. Kurek, M. P., and J. H. Taylor. 1977. Replication of DNA in mammalian chromosomes. IV. Partial characterization of DNA fragments released from replicating DNA of CHO cells. Exp. Cell Res. 104: 7–14.

98. Taylor, J. H. 1977. Increase in DNA replication sites in cells held at the beginning of S phase. Chromosoma 62: 291–300.

99. Taylor, J. H. 1978. Control of the initiation of DNA replication in mammalian cells. In DNA Synthesis: Present and Future. I. Molineux and M. Kohiyama, eds. Plenum, New York. pp. 143–159.

100. Guy, A. L., and J. H. Taylor. 1978. Actinomycin D inhibits initiation of DNA replication in mammalian cells. Proc. Nat. Acad. Sci. USA 75: 6088–6092.

101. Pollock, J. M., Jr., M. Swihart, and J. H. Taylor. 1978. Methylation of DNA in early development: 5-methyl cytosine content of DNA in sea urchin sperm and embryos. Nucleic Acids Res. 5: 4855–4863.

102. Taylor, J. H. 1979. Enzymatic methylation of DNA: patterns and possible regulatory roles. In Molecular Genetics, Part III. J. H. Taylor, ed. Academic Press, New York. pp. 89–115.

103. Taylor, J. H., ed. 1979. Molecular Genetics, Part III. Academic Press, New York. 389 pp.

104. Laughlin, T. J., and J. H. Taylor. 1979. Initiation of DNA replication in chromosomes of Chinese hamster ovary cells. Chromosoma 75: 19–35.

105. Watanabe, S., and J. H. Taylor. 1980. the cloning of an origin of DNA replication of Xenopus laevis. Proc. Natl. Acad. Sci., USA 77: 5292–5296.

106. Taylor, J. H. , and S. Watanabe. 1981. *Eukaryotic origins: studies of a cloned segment from Xenopus laevis and comparisons with human BLUR clones. In* Structure and DNA-Protein Interactions of Replication Origins. ICN-UCLA Symp. Molec. Cell Biol. Vol. XXI. D. S. Ray and C. F. Fox, eds. Academic Press, New York. 597–606.

107. *Sturm, K. S., and J. H. Taylor. 1981. Distribution of 5-methyl cytosine in the DNA of somatic and germline cells from bovine tissues.* Nucleic Acids Res. *9: 4537–4546.*

108. *Taylor, J. H. 1982. Sister chromatid exchange. early experiments with autoradiography. In* Sister Chromatid Exchange. S. Wolff, ed. Wiley, New York.

109. *Chambers, J. C., and J. H. Taylor. 1982. Induction of sister chromatid exchanges by 5-fluorodeoxycytidine: correlation with DNA methylation.* Chromosoma 85: 603–609.

110. *Chambers, J. C., S. Watanabe, and J. H. Taylor. 1982. Dissection of a replication origin in Xenopus DNA.* Proc. Natl. Acad. Sci. USA *79: 5572–5576.*

111. *Taylor, J. H. 1983. DNA replication in mammalian cells. In* Molecular Events in the Replication of Viral and Cellular Genomes. *Y. Becker, ed. Martinus Nijhoff, Netherlands. pp. 115–130.*

112. *Taylor, J. H. 1984.* DNA Methylation and Cell Differentiation. *Springer-Verlag, Vienna. 135 pp.*

113. *Taylor, J. H. 1983. Replicon models for organization and control of chromosome reproduction. In* Genetics: New Frontiers, Proceedings of the XV International Congress of Genetics, New Delhi, December 12–21, 1983. *Oxford & IBH Publishing, New Delhi.*

114. *Taylor, J. H. 1984. Origins of replication and gene regulation.* Mol. Cell. Biochem. *61: 99–109.*

115. *Taylor, J. H. 1984. A brief history of the discovery of sister chromatid exchanges. In* Sister Chromatid Exchanges. R. R. Tice and A. Hollaender, eds. Plenum, New York. pp. 1–9.

116. *Hare, J. T., and J. H. Taylor. 1985. One role for DNA methylation in vertebrate cells is strand discrimination in mismatch repair.* Proc. Natl. Acad. Sci. USA *82: 7350–7354.*

117. *Hare, J. T., and J. H. Taylor. 1985. Methylation directed strand discrimination in mismatch repair. In* Biochemistry and Biology of DNA Methylation. A. Razin and G. L. Cantoni, eds. Liss, New York. pp. 37–44.

118. Riggs, C. D., and J. H. Taylor. 1987. *Sequence organization and developmentally regulated transcription of a family of repetitive DNA sequences of Xenopus laevis.* Nucleic Acids Research *15: 9551-9565.*

119. Taylor, J. H., and J. T. Hare. 1988. *Hemi-methylation dictates strand selection in repair of G/T and A/C mismatches in SV40.* Gene *74: 159–161*

120. Taylor J. H., 1987. *Replication of DNA in eukaryotic chromosomes. In* Results and Problems in Differentiation 14: Structure and Function of Eucaryotic Chromosomes. *W. Hennig, ed, Springer-Verlag, Berlin Heidelberg. pp173-197.*

120. Hare, J. T., and J. H. Taylor. 1988. *Bias in the selection of template strand in mismatch repair of vertebrate cells. Cold Spring Harbor Laboratories meetings, Intermediates in Genetic Research.*

121. Hare, J. T., and J. H. Taylor. 1989. *Cytosine methylation influences the repair of T/G and A/C mismatches in eukaryotic DNA.* Cell Biophys. *15: 29–40.*

122. Taylor, J. H. 1989. *DNA synthesis in chromosomes: implications of early experiments.* BioEssays *10: 121–124.*

123. Taylor, J. H. 1990. *Chromosome reproduction: units of DNA segregation.* BioEssays *12: 289–296.*

124. Taylor, J. H. 1991. *My favorite cells with large chromosomes.* BioEssays *13: 479–487.*

125. Taylor, J. H. 1997. *Tritium-labeled thymidine and early insights into DNA replication and chromosome structure.* Trends in Biochemical Sciences *19: 479-487.*